THE
Technological
Fix

Hagley Perspectives on Business and Culture

Philip Scranton and Roger Horowitz, Editors

Beauty and Business
Commerce, Gender, and Culture in Modern America
Edited by Philip Scranton

Boys and Their Toys
Masculinity, Technology, and Class in America
Edited by Roger Horowitz

Food Nations
Selling Taste in Consumer Societies
Edited by Warren Belasco and Philip Scranton

Commodifying Everything
Relationships of the Market
Edited by Susan Strassera

THE Technological Fix

How People Use Technology To Create and Solve Problems

edited by
Lisa Rosner

Routledge
New York • London

Published in 2004 by
Routledge
270 Madison Avenue
New York, NY 10016
www.routledge-ny.com

Published in Great Britain by
Routledge
2 Park Square
Milton Park, Abingdon
Oxon OX14 4RN U.K.
www.routledge.co.uk

Ceruzzi, Paul. *A History of Modern Computing*, 2nd edition (2003), 319–331. © MIT Press. Reprinted with permission of the MIT Press.

Onno de Wit, Johannes Cornelis Maria van den Ende, Johan Schot, and Ellen van Oost. "Innovative Junctions: Office Technologies in the Netherlands 1880-1980." *Technology and Culture* 43:1 (2002), 50-72. © Society for the History of Technology. Reprinted with permission of The Johns Hopkins University Press.

Printed in the United States of America on acid-free paper.

10 9   8 7 6 5 4 3 2 1

Library of Congress Cataloging-in-Publication Data

The Technological Fix / edited by Lisa Rosner.
    p. cm. -- (Hagley perspectives on business and culture ; v. 6)
    ISBN 0-415-94710-3 (hardcover) -- ISBN 0-415-94711-1 (pbk.) 1. Technology--Social aspects--Case studies. 2. Technological innovations--Case studies. I. Rosner, Lisa. II. Series

    T14.5T3955 2004
    303.48′3—dc22   2003025463

Table of Contents

Introduction

Lisa Rosner

The term *technological fix* is ubiquitous: it is found everywhere in commentaries on technology, whether its past, its present, or its future. Perhaps that is why the phrase is so hard to define. A literal rendering of the words would imply a fix produced by technology, but no one uses it that way. Instead, it has become a dismissive phrase, most often used to describe a quick, cheap fix using inappropriate technology that creates more problems than it solves. But how has this happened? How have the connotations of the words — the tone of voice in which they are expressed — come to overwhelm their denotations? What makes a technological fix not really a fix at all? And under what circumstances can a *mere* technological fix become a true technological solution? This volume, an outgrowth of a conference at the Hagley Museum and Library, attempts to answer these and other questions by examining the technological fix in a variety of historical settings: in medicine, in food production and distribution, in the environment, and in business.

The phrase *technological fix* first gained widespread use in the late 1960s, in the wake of growing concern over the role technology played in social, political, and economic problems. The nuclear scientist Alvin Weinberg may have been the first to use the term in his 1967 book *Reflections on Big Science*. Certainly he was its most enthusiastic promoter. For in the 1960s Weinberg was, as he later admitted, "king of the technological optimists," who saw his own particular technology, nuclear energy, "as a great technical boon for humanity. Eliminating the Malthusian crises, making deserts bloom, burning the rocks — all these seemed to be in our grasp."[1] In that spirit of heady enthusiasm, Weinberg advocated "'cheap technological fixes' that afford shortcuts to resolution of social problems." One of his exemplary fixes was to use the intrauterine device (IUD) as a solution to the problem of birth control, because it did not require individual motivation and thus "provides a promising technological path to the achievement of birth control without having first to solve the infinitely more difficult problem of strongly motivating people to have fewer children." Another was to use nuclear energy to reduce air pollution because his own economic projections showed that it would be cheaper and cleaner than fossil fuels, and "since it is cheaper, people will use it, not because a law requires them to, but because it is to their individual economic advantage to do so."[2] With hindsight, we know that neither the IUD nor nuclear energy can be regarded as either cheap or a fix, however technological they may be. Even at the time, Weinberg's remarks provoked what can charitably be called skepticism. After one talk on the low cost blessings of nuclear power — "the ultimate technological fix that would forever eliminate quarrels over scarce raw materials" — he overheard murmurs of "charlatanism" and "gross overoptimism" as he walked down the corridors.[3]

In Weinberg's eyes, technological fixes were everywhere, as implied by the title of his 1994 memoir, *The First Nuclear Era: The Life and Times of a Technological Fixer*. Though he admitted by then that his cost projections for nuclear energy had been completely wrong, he continued to think of technology as a beneficent force, and the "cheap technological fix" as a solution to social issues.[4] The hydrogen bomb, he argued, had acted as a "stabilizer of relations between the US and USSR," thus serving as a technological fix for preventing World War III. Television and air conditioning, he suggested, could be the technological fix for urban unrest, since it "would keep people off the streets and reasonably comfortable on hot, muggy summer nights." And there would, he claimed, be "a rebirth of nuclear energy, a second nuclear era in which the full potential of the technology will be realized." True, there were critics who pointed out that "technological fixes, because they attack symptoms but don't root out causes, have unforeseen and deleterious side effects that may be worse than the social problem they were intended to solve." But, Weinberg rejoined, social fixes also have their "deleterious and unforeseen consequences," citing Marxism as "the best example of a social fix that was supposed to root out the causes of economic and social injustice, but has caused untold human suffering."[5]

By the time the memoir was published in the 1990s, few readers would have agreed with Weinberg's reasoning on either technology or social issues, despite his many achievements as a scientist. His proposed fixes are too offhand and ill-considered, and his arguments appear to be debating tactics rather than serious discussion of important problems. Even in the late 1960s, his unbounded optimism produced no shortage of rebuttals. In a comment on education for engineers, published in *Technology and Culture* in 1969, John G. Burke argued against a limited technical education, noting that conceiving of social issues in narrow engineering terms would only make them worse. "Although there is no doubt that some of our problems can be solved by sophisticated engineering methods," he wrote,

"I think it is erroneous to insist that all problems are capable of a technological fix." Responding specifically to Weinberg's enthusiasm for the IUD, he continued, "Technological means may be available to solve the population problem, for example, but it has become quite obvious in recent years that sociological and psychological methods will be crucial in effecting a real solution."[6] Clearly in Burke's view, a fix based primarily on the deployment of technology was not necessarily a bad thing: some problems, he noted, could be solved by it. A technological fix, in this sense, was simply a useful technological innovation. But there was a whole class of problems that simply could not be resolved through high technology, however sophisticated the design or however well versed in science and mathematics the engineers. Technology could not fix them, and might distract attention from the true solutions. As Burke used the term, a technological fix was neither good nor bad in itself, but rather appropriate or inappropriate to the task at hand.

By 1970, according to the second edition of the *Oxford English Dictionary*, technological fix was already being used in its wholly negative or ironic sense. The *OED* cites a 1970 book review in the magazine *Nature*, written by Michael Gibbons. Gibbons was reviewing Rene Dubos' *Reason Awake: Science for Man*, a book Gibbons felt reflected the "widespread social discontent about science" that began in the 1960s.[7] For Dubos, the "unbridled development of technology" was largely responsible for the threat to the time-honored, delicate relationship between man and his environment. As Gibbons summarized it, "the ecological balance, according to Dubos, has been evolving over centuries. So, too, has man; his evolution, both mental and physical, is deeply involved with the development of his physical surroundings. Now, with the aid of science and technology, man has so profoundly altered the environment that his physical and mental make-up may no longer be able to cope with the rapid rate of change." Gibbons expressed some doubts over the efficacy of Dubos's "ecological solution" to the ecological problem, contriving somehow to limit growth so as to keep in balance the "myriads of conflicting forces" acting on the future. But he noted with approval the development of an ecological, holistic concept of human society and its approach to social ills, noting its improvement over "the one dimensional 'technological fixes' that society has so far provided to solve its problems."[8] So by the early 1970s, the technological fix was seen as partial, ineffective, unsuccessful, threatening; one-sided as opposed to holistic; mechanical as opposed to ecological.

And yet a paradox remained imbedded in this usage, as Gibbons also noted: the paradox of disillusion with science and technology "in a civilization dedicated to the advancement of knowledge."[9] The paradox continues into the present, and is highlighted in this volume, for we are all, in the twenty-first century, even more dedicated to the advancement of technology than in the 1960s and 1970s. Any issue of *Scientific American* — any issue of any newspaper — reveals hundreds of technological products applied to the solution of hundreds of technological problems. How can we be so enamored with technology and distrust it so much at the same time? As the contributors to this volume show, the answer to this question leads to the heart of some of the central issues in the history of technology. How are problems identified? Who gets to suggest solutions? Are the solutions long-term strategies or merely quick and dirty fixes? Are there differing pressures put on different actors — engineers versus industrialists, or surgeons versus patients — in proposing technological fixes? Can we extend the notion of a technological fix to mean the technical education of the people who implement them? Can there be a technological fix that was held to work in its own day, even though we no longer believe in the efficacy of the technology it employed? Can there be such a thing as a truly successful technological fix, and if so, who gets to define success?

The first set of papers looks at fixing the body. The human body, to many, appears to be the ultimate holistic environment, a complex network of interacting systems. Yet simple fixes, like setting bones, have been available for thousands of years, requiring only a quick, manual operation to allow the body to heal itself. As modern medicine has itself become ever more complex, relying on substantial research in science and technology, the appeal of more intensive — and costly — fixes has increased as well. Persistent metaphors of the body as machine have encouraged these ideas. If the heart is a pump, then surely replacing it is only a matter of designing a body-friendly artificial pump. If the nerve impulses are electrical in nature, then surely electricity can be an effective treatment for nervous disorders. The body-machine continuum has been a fruitful subject for ruminations from science fiction authors. In John Varley's short story "In the Bowl," space-traveling tourists simply acquire the additional body parts or functions necessary to adjust to the conditions found on the planets they visit. Extra arms, eyes capable of infrared vision, and higher capacity lungs can all be purchased and installed as easily as a real-world tourist might buy new sunglasses or other travel accessories. Indeed, so routine is the tech fix to the hazards of space travel that body parts can be repaired and replaced by a mechanically minded eleven-year-old. And it goes almost without saying that people in Varley's world can live much longer than his readers, since body parts, when worn out, can always be replaced.[10] This vision of mechanical repair to the organic body is carried one step further in Stanislav Lem's collection of stories, *The Cyberiad*, which takes place in a universe where squishy biological creatures (like us) have been replaced by cybermachines who construct other machines.[11]

That the body is not just another machine is clearly demonstrated by the three papers in this section. Shelley McKellar examines the development of the artificial heart, a costly and controversial fix to the problem of heart disease. Heart disease was the number one cause of death in the United States throughout the twentieth century, and though it can in some cases be brought under control by lifestyle changes, it cannot be cured. In the post–World War II era, dominated by large-scale engineering projects, the development of an artificial heart seemed a rational and reachable goal, at least to the surgeons and medical technologists who intended to use it. McKellar addresses the questions of "who benefits" and "successful by whose criteria" in her account, examining the funding, the medical personnel, the patients, and the public in the creation of the early artificial hearts. Alarmingly, the group whose bodies were most directly associated with the heart — the early recipients — had the least to say about its use, though later artificial heart patients have been able to reap greater benefits from the technology. This is a technological fix that truly lives up to Gibbons's definition, for it was one-sided and partial, and under the life-and-death conditions of heart surgery, made patients worse rather than better. Still, any medical technology that saves or prolongs life will appeal to dangerously ill patients, and even a technological fix may be more appealing than no fix at all.

Carolyn de la Peña's article looks at a medical fix from the 1920s, an electrical device called the I-ON-A-CO electrical belt, sold and enthusiastically promoted by the Los Angeles real estate entrepreneur Henry Wilshire. The I-ON-A-CO was an oddity, an electrotherapeutic device developed at least ten years after the heyday of medical electricity. De la Peña analyzes its use as a fix in a number of ways. For Wilshire, and indeed for his sales force, it was a profit-generating fix for precarious finances, as well as an empowerment fix providing the opportunity to be valued. For purchasers it was a fix for both body and mind, providing relief from symptoms and a comforting sense of modernity. Wilshire was as incurably optimistic about his device as Weinberg was about nuclear power, and exploring the reasons why Wilshire's technological fix attracted adherents may help to explain why Weinberg's did as well.

The final article in this section, Jim Tobias's examination of debate within the field of technology for those with disabilities, suggests that the final arbiter of whether something is fixed or not is not the designer or promoter but rather the user, and that users themselves may disagree in their evaluation. Tobias discusses the difference between Assistive Technology (AT), technology designed specifically for people with disabilities, and Universal Design (UD), technology that assists people with disabilities but can be used by anyone. These two concepts are influenced by two entirely different views of what is being fixed. In Assistive Technology the assumption is that the disabled person requires the fix, in the form of specialized technology, to function in mainstream society. In Universal Design, in contrast, the assumption is that, as with other technology, there is a market need: it is the world itself that needs a fix, which can be used by all people. In Tobias's discussion of UD, we find ourselves back in the world of John Varley's short science fiction story in which the response to the obvious fact that human organisms cannot instantly evolve to fit their environment is to resort to technology.

In the first section of this book, the technological fixes discussed were proposed in response to physiological conditions, like heart disease or disabilities, inherent in being human, though of course the specific response depended on the unique historical situation. The next section, "Fixing Food," generates additional nuances regarding the meaning of technological fix, for the three fixes were all responses to prior technological change. People have been fixing food, in the sense of altering it to suit their needs, ever since the development of agriculture. The production of cultivated maize from its wild ancestor, the domestication of cattle, the invention of the varieties of plowing implements, were all applications of technology to the complex problem of providing food for human societies. In a real sense no food production is ever completely natural, if we take natural to mean "naturally occurring with no further processing." Even some nonhuman species use tools to obtain food: the use by some species of crows of sticks to get grubs is an effective technological solution. But by the nineteenth century, many of the tasks involved in making and serving food had become heavily industrialized, and these industrial processes generated their own problems in both production and equitable and efficacious distribution. Since it was technological advances that had created the problems, it seemed reasonable to look to technological advances to solve them. Perhaps for this reason, the first two articles in this section explore technologies that historically appeared to be real solutions, not mere fixes.

Michael Ackerman looks at the enrichment of white bread with essential vitamins, a policy advocated by nutritionists and implemented by governments from the mid-twentieth century until today. White bread has had a long history as an upper-class commodity, making it the food of choice for all income levels by the end of the nineteenth century. But the refining process necessary to produce white flour eliminated many essential nutrients. As nutritionists became aware of this problem, caused by processing food, they were divided over the best solution. One approach was to educate consumers to prefer whole-grain, dark breads on the grounds that "nature knew best"; the other was the technological fix of adding nutrients back into the white flour during the production process. What Ackerman's article clearly shows is that one person's successful solution is another person's partial quick fix. Even professional experts like nutritionists can have opposing viewpoints on the success of proposed solutions; something to be borne in mind by potential clients and consumers.

Shane Hamilton studies the developments in transportation of frozen foods as a solution to several problems. The most broad-based issue was the basic economics of food production and consumption. By the mid-twentieth century, American farmers were too successful in producing food for their own good. Increased output meant lower prices, which made it uneconomic to be a farmer. But raising prices artificially was not the answer, because consumers in the ever-growing cities demanded low-cost food. The U.S. Department of Agriculture sought the answer in cutting costs of transportation. In one important area, transportation of frozen food, trucking took over from railroads because it was faster, more flexible, and more adaptable to the needs of specific commercial foodstuffs. Trucking culture, seldom paired with high tech engineering in the American consciousness, nonetheless emerges in Hamilton's account as an outgrowth of technological solutions to complex problems.

Warren Belasco's article is on food fantasies rather than realities. For this reason it highlights an important aspect of technological fixes not covered in Gibbons's definition: many are imaginary or visionary solutions to complex problems. Two pervasive problems in the history of food are the difficulty of feeding an ever-growing population and the often inequitable distribution of labor in getting food on the table, with women bearing the brunt of that labor. In a technologically sophisticated world, Belasco shows, it is easy to imagine, if not carry out, high tech solutions from pills to algae burgers. And when products like Slimfast and eggs-on-a-stick are marketed as nutritious time-savers, can we be sure that the distance between imagination and reality is so great?

To paraphrase Mark Twain, everybody talks about the environment, but nobody does anything about it. The third section takes a look at what "doing something" about the environment might mean. People have always fixed the environment as they have food to serve their own needs, and competing visions of the most useful environment have provoked conflict as surely as competing visions of the best form of government. But the environment, like the human body, is a complex system and does not lend itself easily to fixing. This has been made very clear to us when ecological systems are upset, whether by importing rabbits into Australia or all the birds in Shakespeare into New York's Central Park. It is also made clear by the persistence of pollution problems despite centuries of attempts to solve them. Since the growth in awareness of environmental degradation was partly responsible for the negative view of technology that first inspired Gibbons's use of the phrase, we should not be surprised that the articles in this section look unfavorably on the proposed technological fixes. Indeed, all three articles raise the question of whether it is our reliance on technological fixes that throws the system still more out of balance and has prevented us from finding true solutions to environmental issues.

Environmental technological fixes, like others in this volume, were invoked to solve problems that technology itself, whether industrial or military, had caused. Frank Uekoetter gives us a case study in the history of air pollution control in his investigation of the search for a solution to the problem of coal smoke from furnaces in the early twentieth century in Germany and the United States, two leading industrial nations. Since coal was the main source of energy, and furnaces were, generally, poorly designed, air pollution was a problem in every major city. This technological problem resulted in concomitant demand for a technological fix, one that inventors and purveyors of "smoke consuming devices" were anxious to supply. Alas, as Uekoetter describes, the solution was not so simple. Some inventions were simply frauds. Others worked to some extent, but the extravagant claims made for them discredited even honest efforts. The result was an emphasis on quick, cheap fixes. But Uekoetter's paper can also add another nuance to the definition of technological fix. In both the United States and Germany, the boiler room became the focus of antipollution activity, but in different ways. In the United States the consensus was that engineers should solve the problem. In Germany, in contrast, it was the firemen who ran the boiler room who became the focus of attention. We see in this instance people as a kind of fix, a technological one insofar as they rely on technological expertise to solve the problem.

Timothy LeCain's article adds to the complexity of the definition of technological fix by raising the issue of whether a true fix is possible, or whether *eco-efficiency* is a kind of zero-sum game in which one player can only win at the expense of another. LeCain's players are primarily businesses and the engineers who work for them; the stakes are profits and the avoidance of law suits. Their strategies are three types of technological fixes. In the transformational techno-fix, a polluting by-product of an industrial process is transformed through the "miracle of modern technology" into a useful and saleable commodity. Techniques for this are closely linked to the relocational techno-fix, in which the commodity is actually sold. The problem in both these types of techno-fixes comes from the proverbial law of unintended consequences, which can lead to LeCain's third category, the delayed techno-fix, in which environmental degradation is simply shifted a generation or two into the future.

James Fleming's paper gets us back to Mark Twain's original topic, the weather. Military implications of predicting and controlling the weather have been clear at least since William of Normandy prayed for favorable winds to cross the English Channel in 1066. Just after World War I, Viehelm Bjerknes and his colleagues came up with the designation *polar front* by explicit analogy with military fronts; during World War II his son Jacob (Jack) organized a training school for Air Force weather officers. As Fleming describes, during the Cold War emphasis shifted from knowledge of the weather to techniques for influencing it. Chief among these was cloud seeding, adding a small amount of a *nucleating* agent (usually dry ice) to induce rain, fog, and snow. Like Uekoetter's inventors of smoke-abatement devices, advocates of cloud seeding stressed its unlimited promise as a quick, effective technological fix: it would allow pilots to dissipate cloud or fog covering bomb sites and landing areas, tame unfavorable winds, and turn the atmosphere itself into a weapon. To countries engaged in a nuclear arms race, shifting control of weather from divine to human hands seemed the obvious next step. If there is to be a fix at all, however, it may be the same sort that we observed in Uekoetter's paper, of people rather than technology. Fleming, like Uekoetter, charts the rise of educated experts, in his case civilian meteorologists funded by the National Aeronautics and Space Administration (NASA) and the National Science Foundation (NSF) rather than the Department of Defense. They felt weather modification only brought negative publicity, interfered with free exchange of scientific data, and distracted attention from serious environmental concerns.

The final section looks at fixing business, where the problems are less apt to involve major social upheaval and are required to reflect a concern for profit. Technology for business, especially the office environment, has been developed to serve diverse, and sometimes diverging, goals. The ones that most readily spring to mind are increased productivity, efficiency, and profitability, and certainly technologies from telephones to typewriters, from data files to the Internet, have helped achieve those goals. But there are others, including increased access to information, or, to take the other side, increased privacy; they also include greater diversity in the workforce or more opportunities for those already there. Because many office tasks need to be done over and over for the office to function successfully, there is a built-in market niche. And because businesses can deduct the cost of many technologies from taxes, there is a built-in incentive for using them; or put more precisely, the built-in disincentive, the cost of new equipment, is not as strong.

The two articles in this section return us to at least a relatively positive usage of the term technological fix, because in both cases there is a genuine technological problem that is solved by technology itself. Though Onno de Wit, Jan Van den Ende, Johan Schot, and Ellen van Oost do not use the term technological fix in their analysis of office technology in the Netherlands between 1880 and 1980, it is implied in their analysis of what they call "innovation junctions." These are places, such as factories, hospitals, and in this article, offices, in which users bring together different technologies for a set of complex uses, leading to interaction among users and among technologies. The result is a distinctive pattern of innovation. In the first phase, new office machines, like typewriters, were introduced to allow specific sets of users (managers, secretaries, clerks) to perform specific tasks. Once they were in place, it made it more likely that other complementary technologies would be introduced, and that these technologies would spread to new groups of users. In the second phase, management experts and managers themselves advocated organization systems that often relied on multifunctional office machines, like addressing machines, which could combine what had once been separate tasks. In the third phase, the electronic computer, with its tremendous capacity for centralized office integration, increasingly came to dominate. This article suggests that technological fixes need not come singly; instead, certain kinds of environments may create the conditions for multiple, and interacting, technological solutions.

Paul Ceruzzi closes the volume with a very recent technological fix, the development of technology for computer-to-computer communication that makes the modern Internet possible. Engineers of the early 1990s were confronted with the difficulty of getting computers of different types, using different operating systems, to communicate with one another. Though this problem existed from the early period of personal computers, when each manufacturer developed its own proprietary software, it became especially acute as more and more computers were hooked up to the World Wide Web. The technical problems were solved with astonishing ease in the enhancement of hardware and the development of Web browsers and Java. But along the way, the other great fix the early advocates of the Internet had hoped for, more democratic end-user access to information of all sorts, was abandoned. As Ceruzzi shows us, even our most recent technological fix, like the others recounted in this volume, ended up being only a partial solution to a complex problem.

The study of the technological fix, then, takes us from visions, to trials, to partial or full solutions of social and economic problems, both simple and complex. It looks at people who design the solutions and people who are the beneficiaries or victims of the design. In other words, it takes us through the whole range of technology itself. The articles in this book reveal much about the way people in the industrialized world think about and use technology: they serve as fascinating case studies and as stimuli to additional research.

The contributors and editor would like to thank all those who made this volume possible, especially Roger Horowitz and Phil Scranton, series editors; Ritchie Garrison, Arwen Mohun, Janet Davidson, commentators at the Technological Fix conference in October 2002; and Karen Wolny, William Germano, and Jaclyn Bergeron at Routledge, and Marsha Hecht at CRC Press.

Notes

1. Alvin Weinberg, *The First Nuclear Era. The Life and Times of a Technological Fixer* (New York: AIP Press, 1994), p. 151.
2. Weinberg, *First Nuclear Era* , p. 150.
3. Alvin Weinberg, *Reflections on Big Science* (Cambridge, MA: MIT Press, 1967).
4. Weinberg, *First Nuclear Era*, p. 150.
5. Weinberg, *Reflections*, p. 158.
6. Weinberg, *First Nuclear Era*, p. 151.
7. John G. Burke, "Comment: Let's Be Sure Technology Is for Man," *Technology and Culture* 10.1 (January 1969): 13.
8. Useful surveys of attitudes toward science and technology in the early 1970s can be found in Donald W. Shriver Jr., "Man and His Machines: Four Angles of Vision," *Technology and Culture* 13.4 (October 1972): 531–555 and Paul T. Durbin, "Technology and Values: A Philosopher's Perspective," *Technology and Culture* 13.4 (October 1972): 556–576.
9. Michael Gibbons, "Scientists and Society," *Nature* 228 (October 24, 1970): 387.
10. Ibid.
11. John Varley, "In the Bowl," in Robert Silverberg and Martin Greenberg, eds., *Great Science Fiction of the 20th Century* (New York: Avenel Books, 1987).
12. Stanislav Lem, *The Cyberiad* (New York: Harvest Books, 2002).

I
Fixing Bodies

1

Artificial Hearts: A Technological Fix More Monstrous Than Miraculous?[1]

SHELLEY MCKELLAR

The twentieth century was a period of dramatic surgical innovations, during which surgeons increasingly performed ever greater interventionist reparative operations. Surgery as a therapy expanded from simply cutting out disease (such as taking out tumors) to repairing damaged internal structures (such as vascular repairs or grafts) to replacing body parts (such as organ and tissue transplantation). The growth in surgical scope and success was directly linked to the discipline's ability to manage pain, bleeding, and infection through such innovations as anesthesia, heparin, and penicillin.[2] In the post-World War II period, surgeons and engineers began to work more closely to explore the possibilities of technology in treating disease, specifically inventing devices that assisted organ functions or replaced body parts altogether. Surgical procedures such as hip replacements, pacemaker implants, and numerous prosthetic operations became routine. The era of rebuilding bodies — and "fixing" the ailing body — was underway, ranging from artificial limbs to internal organs.[3]

In the case of the human heart, could a mechanical device truly replace it? If so, what would it be made of, be powered by, and/or look like? What materials were most compatible with body fluids? Should such a device be pneumatic, electric, or even nuclear-powered? Would it be a machine outside the body or would it be an implanted device? In the postwar period, scientists, engineers, and clinicians sought to overcome many of these issues, and numerous devices began to emerge to pace, to replace, to bypass, and to assist pumping for the heart. Such technological innovations would have enormous utility for a society plagued by various heart afflictions. To what extent was the total artificial heart conceptualized as a technological fix for the problem of cardiovascular disease, the leading cause of death in the United States? This paper explores some of the scientific and public responses to the total artificial heart as a viable treatment to fight end-stage heart disease, and highlights the role of the state, specifically the artificial heart program of the National Heart, Lung, and Blood Institute (one of the institutes of the National Institutes of Health [NIH] in Bethesda, Maryland), in providing much needed funding and encouraging an increasing number of experts to pursue this line of research toward finding a mechanical solution to this biological problem.

Heart Disease and Early Investigations

Heart disease is a prominent cause of death in affluent nations. In the United States it has been the number one killer every year since 1900 (except 1918). In fact, cardiovascular disease claims almost as many American lives each year as the next seven leading causes of death combined.[4] Heart failure affects an estimated 4.7 million Americans, with 550,000 new cases diagnosed annually and annual cost estimates ranging from $10 billion to $40 billion.[5] Heart disease may be controlled through eating the right foods, exercising, and not smoking. Nevertheless, a technological fix to address the problem of cardiovascular disease held much greater appeal than health promotion or disease prevention programs during an era of technological achievements and optimism.

In the United States, the public's first introduction to the term "artificial heart" came in 1935. At the Rockefeller Institute for Medical Research in New York, Alexis Carrel and Charles Lindbergh built a perfusion pump — a glass apparatus that permitted an animal organ to be sustained and observed by scientists (see Figure 1.1). The idea was to maintain, in a living state, a portion of the body in order to study its functions. Instead of merely guessing at changes that took place in an organ, scientists hoped to identify significant details about heart disease, kidney disorders, or even contagious ailments.

Alexis Carrel was a Nobel prize–winning French scientist and surgeon. Charles Lindbergh was the celebrated American aviator who flew solo over the Atlantic Ocean in 1927. Together they published their work in several articles, but it was their 1935 article in *Science* magazine that garnered the greatest attention within the scientific community. In that article, Carrel and Lindbergh stated that in their 26 experiments performed with the latest model of their device, organs such as glands, spleens, livers, hearts, and kidneys of various animals (predominantly chickens and cats) had been kept alive. They referred to their device as "an apparatus for the culture of whole organs."[6]

Fig. 1.1 The Lindberg-Carrel perfusion pump, 1930s, designed to keep an isolated organ supplied with enough blood and oxygen to maintain its vitality for an indefinite period of time outside the body (Courtesy of the Smithsonian Institution's Division of Science, Medicine & Society).

New York Times science reporter William L. Laurence referred to it as an artificial heart, and the term stuck.[7] Clearly this was a misnomer from the standpoint of today's definition. The Carrel-Lindbergh device was never intended to replace the heart, either outside or inside the body, in an effort to save the heart failure patient. It was, in fact, a perfusion pump designed to keep one isolated organ (a thyroid or spleen, for example) supplied with enough blood and oxygen to maintain its vitality for an indefinite period of time, outside of the body. But reporters and the public latched onto the term artificial heart nonetheless because the device did pump blood and oxygen to the isolated organ to keep it alive. Arguably one might state that it was a very early step toward today's mechanical hearts. The Carrel-Lindbergh artificial heart did make it possible for the first time to keep an organ alive outside the body.

The public's interest and imagination was stirred; they were excited and awed by this type of medical science. Hundreds, even thousands, of people attended Carrel's lecture series, at which he predicted that the artificial heart would lead to the day when parts of humans, or even entire bodies, could be placed in suspended animation for long periods. His ultimate dream was to maintain an inventory of living organs away from the body, ready for transplant. He outlined his argument in his 1938 book, *The Culture of Organs.*[8] *Time* magazine in 1938 ran a cover story on the pump and reported that Carrel and Lindbergh "are going to transform medicine from a fine art into an exact science."[9]

One might credit Lindbergh as one of the first people to think seriously about ways to temporarily replace the function of the heart and lung. He came up with an early crude pump-oxygenator. Lindbergh did not pursue education beyond high school, yet as one historian of technology noted, "As preposterous as it seems for a pilot to be masquerading as medical researcher, the fact is that Lindbergh brought the new field of artificial organs exactly what it needed: practical engineering."[10] Medicine and engineering were both needed in developing artificial organs, and that was what Carrel and Lindbergh had in fact done by teaming up together. Together, they demonstrated that a machine that works very much like a heart could sustain life. The more promising artificial heart research emerged after World War II but had very little to do with Carrel's misnamed artificial heart.

During the late 1940s and into the 1950s, a handful of researchers conducted work on mechanical circulatory support systems, including heart-lung bypass machines. In regard to specific artificial heart research, Dr. Willem Kolff worked on a variety of prototypes in his lab at the Cleveland Clinic. The inventor of the artificial kidney machine, Kolff would spend the rest of his career leading research teams exploring artificial eyes, ears, arms, lungs, and hearts.[11] In December 1957, Dr. Tetsuzo Akutsu and Dr. Kolff implanted the first artificial heart in the Western world in a dog, and maintained its circulation for 90 minutes.[12] It was an air-driven artificial heart made of polyvinyl chloride. It marked the beginning of years of experiments by Kolff of various types of artificial hearts, including pendulum, atomic energy, and electromagnetic hearts.[13] His artificial hearts varied greatly in form and method of operation. At this early stage of experimental hearts, Kolff and his associates identified basic problems and challenges in four main areas: (1) the elementary components and material of which the artificial heart was to be made; (2) the source of energy; (3) the driving mechanism; and (4) the regulating devices.[14] The involvement of the state would bring much needed money and encourage additional experts to pursue this line of research, with expectation that the problems associated with the artificial heart would soon be overcome.

The Role of the State: The Artificial Heart Program

In 1964 the National Heart Institute (later renamed the National Heart, Lung, and Blood Institute) of the National Institutes of Health formally established an artificial heart program of publicly funded research and development. Their objective was to develop a range of devices to assist the failing heart and to rehabilitate patients with heart failure. The list of devices included emergency devices, temporary devices and instrumentation, short-term circulatory assist devices, long-term assist devices, and long-term replacement devices — the most prominent being an implantable mechanical heart or total artificial heart.[15] The artificial heart program would fund such work through research contracts, supporting a number of academic and industry-based research and development teams. Congress told the National Heart Institute to pursue development of the artificial heart "with a sense of urgency" and to encourage as many researchers as possible "to develop devices and materials [to this end] and to explore new avenues of approach."[16] It was an era of unbounded enthusiasm for accepting new challenges, such as space travel to the moon.

Key to understanding the establishment of the artificial heart program is the recognition that most of those agencies and individuals who supported the new program perceived the problems of development to be largely engineering in nature, and thus amenable to the same general approach as that being concurrently used in the U.S. space program.[17] It was approached on a systems development basis or the separate development of each component such as energy systems, pump designs, and blood interface materials toward later integration into a clinically viable cardiac device.[18] This was based on the presumption that the necessary science base and related technologies from this and other fields could be readily adopted for artificial heart development.[19]

In an era of medical breakthroughs and technological achievements, the public applauded bold, new procedures and undertakings. The artificial heart program benefited from both a receptive public environment as well as a pattern of publicly funded, successful technology programs. Historically, the state had encouraged and paid for numerous large-scale research and development operations to achieve rapid progress on technological programs, including the Panama Canal, the Colorado River dam, and the atomic bomb. As one historian of technology noted, many Americans believed in their ability to solve complex technological problems simply by gathering the best minds, breaking up the task into manageable segments, coordinating the subunits from the top, and spending whatever it took to get the job done.[20] Public confidence in the ability of medical researchers was high due to recent achievements in improving and extending life through the artificial kidney machine or dialysis, cardiac surgery innovations including heart bypass and pacemaker implants, as well as the promise of organ transplantation. All of these efforts were supported by grants from the National Institutes of Health, so why not a total artificial heart? The consensus among most investigators was that money and industry's technological expertise were the missing components needed to develop the artificial heart.[21]

Why did they think that building an artificial heart would be this straightforward? It was because they viewed the heart as a double pump, and Americans already knew how to build pumps.[22] As a pump, the heart circulates blood throughout the body. It pushes oxygenated blood from the lungs into the left atrium of the heart and from there into the left ventricle. The left ventricle pumps the blood via the aorta into arteries throughout the body. After oxygen-rich blood has nourished the body's organs and capillaries, the blood flows through veins back to the heart's right atrium and then to the right ventricle. This chamber sends the venous blood to the lungs to be oxygenated, completing a pumping cycle that normally occurs 70 to 90 times per minute.[23]

Today, patients with mild to moderate heart failure are most commonly treated with drug therapy. For patients with severe heart failure, the treatment of choice is heart transplantation. At the time of the creation of the artificial heart program, the first successful human heart transplant had yet to be done. It was still three years off. (In 1967, South African Christiaan Barnard performed the first heart transplant.[24]) Presently, about 2,200 heart transplants are performed each year with thousands of patients on waiting lists for a donor heart. Many other heart patients are deemed unsuitable candidates for transplantation. The artificial heart program was thus launched in the spirit of providing an alternative treatment for end-stage heart disease, acknowledging the limitations of transplantation and drug treatment.[25]

Total artificial hearts are devices that are orthotopically placed in the human body after the removal of the natural heart.[26] Surgeons remove almost all of the natural heart and insert the mechanical heart in its place. In design and operation, the mechanical heart is basically a two-pump chamber with one power source that performs the functions of the ventricles. Power to drive the device is transmitted through the skin. One of the initial goals of the artificial heart program was to develop a fully implantable total artificial heart by Valentine's Day 1970.[27] Many believed this goal to be attainable. Scientists, clinicians, and sponsoring agencies told skeptics to think about the number of lives that could be saved with such a device.

The conceptualization of as well as the rhetoric or words that government administrators, scientists, clinicians, patients, and journalists used to describe and articulate the problem of and solutions to heart disease are an important part of this story. As articulated in the objectives and goals of the artificial heart program, experts in the field aimed to replicate mechanically the function of the human heart. So long as the heart was just a pump, it resembled other inorganic objects that moved fluids around. Conceptually it was an engineering and materials problem.[28] Engineers and scientists took their tasks seriously, however they did not anticipate the body's response to such a complex foreign device nor the difficulties of building a long-term replacement device. Building an artificial heart and warding off death would be much more difficult than building a rocket and putting man on the moon.

The Artificial Heart and Clinical Trials, 1969–1991

There is a large cast of characters who participated in the laboratory development and clinical trials of various models of artificial hearts. The process through which these devices were tested in the laboratory, both in biomaterials testing and in animal trials will not be discussed, although it is an important part of the story. Instead, the more well-known clinical cases will be examined, focusing on the device and to a lesser extent the participating scientists, clinicians, and patients.

The first human case in which a patient was "saved" by an artificial heart took place in 1969 at the Texas Heart Institute. Cardiac surgeon Denton Cooley implanted an experimental total artificial heart — the Liotta total artificial heart — in 47-year-old Haskell Karp (see Figure 1.2). Designed by Domingo Liotta, M.D., it was a pneumatic total artificial heart, which, as a two-chambered device, functioned much like a natural heart with one noticeable exception. It was powered by enormous air pumps outside the body, using hoses to pass through the patient's body wall and into the circulatory system. The intent of implanting this highly experimental mechanical heart in Karp was to keep him alive until a human heart was available for transplantation. Thus it was as a temporary device (a bridge-to-transplant) to which the artificial heart served — and it worked — but it presented a troubling image and scenario for some family members. Karp's wife described her husband's condition this way: "I saw an apparatus going into the arms, the hands, the feet. He could not say anything. I don't think that he was really conscious. I see him lying there breathing, and knowing that within his chest is a man-made implement where there should be a God-given heart."[29] The artificial heart kept Karp alive for approximately 64 hours; at that time he underwent a heart transplant operation. Unfortunately Karp died 32 hours after the transplant operation.[30]

The heart operation was not without controversy. Cooley had performed the surgery without permission from the Baylor Committee on Research Involving Human Beings. There was bad publicity from that, as well as in-house feuding between Cooley and his mentor, Dr. Michael DeBakey, also working on mechanical heart devices at the time. According to DeBakey, the artificial heart was "covertly taken from the Baylor surgical laboratory to St. Luke's Hospital," where Cooley performed the surgery.[31] Technically, it was the first successful total artificial heart bridge-to-transplant operation.

Many scientists and administrators in the field however felt that human implantation was premature and detrimental to the artificial heart program. There were technical problems in the device yet to be worked out, specifically biomaterials and a power supply.[32] Cooley would almost certainly never have received approval if he had sought it in advance of the operation. Cooley told a *Life* magazine reporter that, "It wouldn't have done any good to call them [Baylor Committee on Research Involving Human Beings]. DeBakey runs the committee and they would have automatically turned me down."[33] The case was dismissed as experimental.

Fig. 1.2 Liotta total artificial heart, implanted in a human as bridge-to-transplant in 1969 by cardiac surgeon Denton A. Cooley (Courtesy of the Smithsonian Institution's Division of Science, Medicine & Society.)

An artificial heart research conference was held in 1969, only months after the total artificial heart implant by Cooley. In their published proceedings, program participants stated that the philosophy of the artificial heart program should be to stress rehabilitation rather than mere prolongation of life.[34] Perhaps some participants anticipated the forthcoming debate about the use of the artificial heart as a temporary versus a permanent device, and about the criticisms of half-way technology versus high technology. Was the device intended to be a half-way technology to overcome a disease's impact only after it was fully developed to postpone death, or should its development be aimed at high technology objectives of preventing or curing disease?[35] (This is a difficult debate in which to position artificial heart technology for "relatively few disease entities appear to be susceptible to a high technology, at least in the short term."[36]) Equally significant, emphasis in the artificial heart program shifted from separate development of each component to integrating existing components to produce a fully implantable cardiac assist device for testing in animals. Redirection of the artificial heart program was not unrelated to Cooley's experimental implant case or the first heart transplant cases now underway around the world.[37]

In 1981, Cooley performed a second total artificial heart implant surgery at the Texas Heart Institute. A less crude device, the Akutsu III total artificial heart (named after the designer Tetsuzo Akutsu, M.D.), consisted of fist-size polyurethane globes that connected to the patient's atria, aorta, and pulmonary artery via Dacron connectors. It was pneumatically driven, once again tethering the patient to an air compressor at the bedside. Cooley implanted the Akutsu artificial heart in 36-year-old Willebrordus Meuffels, who was dying of heart failure while awaiting a human heart transplant. The artificial heart kept Meuffels alive for several critical days, at which time a suitable donor heart became available and a transplant operation was performed. Unfortunately, Meuffels died shortly after the transplant due to organ rejection, infection, as well as kidney and lung problems.[38]

Once again Cooley had performed the operation without permission from any regulatory body, notably the Food and Drug Administration, which licensed new medical devices. Cooley explained: "If a man is overboard and someone throws him a life preserver, he's not going to inspect it and see if it has a guarantee."[39] For him, it was about saving a life and he could not be bothered with time-consuming red tape or overly cautious administrators.

Watching these events at Texas Heart Institute closely, another team of researchers and clinicians in Salt Lake City, Utah, began preparing for their own clinical case. In 1982 Dr. William DeVries implanted a Jarvik-7 artificial heart into Barney Clark as a permanent replacement for a natural heart (as opposed to a bridge-to-transplant in both of Cooley's cases). This would be an historical first, if successful. The Jarvik heart, named after designer Robert Jarvik, M.D., emerged from Dr. Kolff's research lab, now based out of the University of Utah. The Utah research and surgical team had planned for months for this implant, going through proper channels to receive approval from both the university hospital's institutional review board and the Food and Drug Administration. Barney Clark, a 61-year-old dentist from Seattle, was dying of heart failure. He signed a lengthy consent form — twice — to undergo the experimental operation. According to his wife, Clark "was astutely aware of his critical physical condition and the highly experimental nature of the implantation and did not expect any great personal miracle; he knew full well that by volunteering, he would be submitting to a totally unpredictable future ..."[40] No one, including his medical team, anticipated the complications that followed the operation.

Like the earlier artificial heart implants, the Jarvik-7 was an air-driven pump, and Clark was bound to an air compressor the size of a washing machine, which powered it. As with the previous patients, there were tubes from the compressor passing through Barney Clark's chest wall, binding him to his bed and causing constant infections. Clark's blood also kept clotting as it passed through the imperfect man-made pump. Clark suffered a number of seizures, strokes, and even a mechanical failure (broken valve) of the heart before he died 112 days after the implantation.[41]

The 1982 procedure was a media event. There were daily media reports scrutinizing and chronicling the details of Barney Clark's 112 days of life with the Jarvik-7. Overall the reporting was optimistic and celebratory. It was a "race to have a heart" — a rescue-oriented surgery — conveying hope and faith in medical technology and technological solutions.[42] Over the 112 days the public witnessed firsthand human experimentation and it was gruesome. They watched Clark experience kidney failure, convulsions, memory lapses, and eventually death. Television interviews with Clark showed a zombie-like appearance with halted, stilted speech. When Clark died, after a questionable period of living with the artificial heart, mixed feelings among journalists and the public were apparent. The medical community, as expected, praised Clark as a "medical pioneer" and proclaimed the artificial heart a success in delivering borrowed days and weeks to Clark and his family.[43]

Hospital administrators at the University of Utah stalled the process for the next implant patient, and to their credit, examined the medical, ethical, and financial issues raised by the Barney Clark case.[44] Months later, DeVries moved to Louisville, Kentucky, to continue his clinical trial. In 1984 and 1985, at the Humana Heart Institute, he implanted Jarvik-7 artificial hearts in three more patients. A Jarvik-7 heart was also implanted in a patient in Sweden. The longest survivor was William Schroeder who lived for a record 620 days (20 months).[45] In all five cases, including Barney Clark, there were serious clinical problems, particularly with strokes. An implantable power supply and nonclotting surfaces also remained key problems.[46]

A wave of artificial heart implants followed worldwide. Only five patients in total received a Jarvik-7 (renamed Symbion) total artificial heart as a permanent circulatory support device. Less known is that another 218 patients were implanted with total artificial hearts as a temporary device between 1985 and 1991.[47] As a bridge-to-transplant, the artificial heart was implanted in patients with heart failure who were deteriorating while on the transplant waiting list. The implant worked for roughly 65 percent of these patients, of whom 50 percent were alive one year after transplant. Eleven models of total artificial hearts were implanted between 1969 and 1991. All were pneumatic in design, thus requiring positioning of drivelines or connections to an external power source, but had different variations of the drive system and monitoring units. The most frequently used TAH model was the Symbion J7-70, a smaller version of the original Jarvik-7. Regardless of model, complications of implantation continued to be infection and hemorrhage, with less incidence of neurologic events (or strokes).[48]

Media attention on the use of the total artificial heart as a temporary device or bridge-to-transplant was minimal compared to its application as a permanent implant by DeVries. The public's fascination with an artificial heart had faded after the Barney Clark case. In 1988 a *New York Times* reporter dubbed the artificial heart "the Dracula of Medical Technology" for "sucking $240 million out of the National Heart, Lung, and Blood Institute."[49] In the public's mind as well as many bioethicists and other academics following the total artificial heart implants, the Jarvik-7 artificial heart had perhaps pushed science too far, too fast. For some, the artificial heart symbolized a technological fix that had turned more monstrous than miraculous and necessitated reassessment. The mechanical biological problems associated with building an artificial heart as well as increased concern as to the ultimate economic, ethical, and social implications of this technology fueled debate in both the scientific and public domains as to whether this line of research should continue.

Conclusion: Mechanical Fixes to Biological Problems

How does one evaluate the artificial heart program? Has this been time, expertise, and money well spent? It is clear that long-term research support by the National Heart, Lung, and Blood Institute was necessary in this area of research as it was beyond the means of independent investigators and their institutions. It is also clear that artificial heart research and development was more complicated than initially predicted, and unlike large-scale technology programs of the past, the artificial heart raised difficult ethical issues. Certainly, investigators gained increased biomedical knowledge surrounding mechanical circulatory support systems.[50] Development costs of technologies were later to be assumed by industry as clinically effective devices moved toward marketing approval.

Important new cardiac technologies aside from total artificial hearts emerged, specifically devices of a temporary and assistive rather than replacement nature. Among the emergency, temporary, or short-term devices that emerged were the intra-aortic balloon pump, the membrane oxygenator, and ventricular assist devices (VADs). VADs as partial artificial hearts have proven effective as both temporary and permanent devices. Increasing data collected on patient experiences with partial artificial hearts show encouraging uses of this technology as bridge-to-transplant, bridge-to-recovery, and lifelong therapy.[51]

Four decades of support to build a mechanical heart indirectly contributed to the debate surrounding preventative versus therapeutic medical initiatives. Many researchers and policy makers argued that medical research funding was better spent on disease etiology study and/or preventative programs rather than applied technology pursuits.[52] Yet there remained a solid core of investigators committed to developing a mechanical alternative for diseased or damaged organs. After the first Jarvik-7 clinical case, Dr. Willem Kolff wrote this to Barney Clark's son: "As to the question whether or not your father's experience and pioneership has advanced the cause of the artificial heart, *there is absolutely no doubt about it.* As far as the question whether or not an artificial heart will eventually be able to restore people to a happy existence, *there has never been any doubt in my mind* about it."[53] [author's italics]

In 1999 scientific reviewers of the artificial heart program concluded that although biological, technological, and socioeconomic concerns persisted, "there are no insoluble problems that nullify the expectation of a clinically effective TAH."[54] Many social scientists queried these findings: even if it could be built, what would be the cost and who would have access? The social, legal, ethical, and economic dimensions of this technology continue to be debated.[55] According to most in the biomedical community, concerns of access, technology assessment, patient decision-making, and quality of life are resolvable issues. There are no other emerging therapies to offer the estimated five to ten thousand heart failure patients who could possibly be well served by a total artificial heart. Scientific reviewers of the artificial heart program concluded that NIH funding should continue.[56]

Support for the development of an artificial heart reflected American values and attitudes regarding the utilization of machines and devices to fight disease, endorsing the technological imperative of twentieth-century medicine. Despite criticisms against the quest for an artificial heart at various times since the establishment of the artificial heart program in 1964, NIH support continued.[57] Government funding was directed to research on total and partial artificial hearts, as short- and long-term devices, and the medical device industry took a larger role in continuing development, banking on forthcoming device refinements to make it a feasible and profitable therapy.

Artificial organs (in particular total artificial hearts) challenge what a society thinks about the limits and potential of mechanical devices as medical therapies. The technology is a mechanical response to a biological set of problems. In the past, technology has made significant contributions to medical therapeutics including hip replacements, dialysis, prosthetics, stents, and more. For an aging baby boom population, the idea that ailing bodies can be fixed or rebuilt from artificial knees to hearts is perhaps reassuring. Yet artificial organ technologies raise contentious issues of illness and treatment, disability and body autonomy, and elicit significant differences of experiences and meaning.

The medical research community appears to be shifting its focus from mechanical to biological solutions for curing disease, the goal being to manipulate the body to heal itself.[58] Perhaps the metaphor of our bodies as machines no longer applies or at the very least needs to recognize limits. Repair and replacement of some damaged body parts, such as a heart, proved to be much more difficult than expected. Humans are not machines, and rather than repair or replacement, bodies need to heal or they die. Healing is the key in a society less enthralled today with technological fixes than a generation ago. Only time will tell to what extent society will continue to support bionic medicine as it debates issues of access, cost, and quality of life, and as it reassesses the power of human intervention to fix things.

Postscript: Twenty-First–Century Artificial Hearts

In 2001 the medical device firm AbioMed received FDA approval to implant their AbioCor total artificial heart device in heart failure patients as a permanent implant (see Figure 1.3). Almost twenty years after Barney Clark, surgeons implanted a smaller, improved artificial heart in 59-year-old Robert Tools in Louisville, Kentucky. The technology was better. One significant improvement was that the device transmitted power transcutaneously through the skin without need for wires or tubes (and thus reduced infection). The mechanical heart was also powered electrically through an implanted titanium and polyurethane plastic battery in the abdominal cavity, not by a large, external air compressor. The implanted battery renewed its power from an energy transfer coil through the skin or transcutaneous energy transmission (TET). The AbioCor artificial heart also offered mobility to the patient. Whereas Barney Clark, William Schroeder, and other 1980s heart implant patients never left hospital surroundings, Robert Tools made well-publicized day trips to nearby restaurants and parks.

When interviewed in August 2001, Tools responded to how well he felt, beaming for the cameras and telling reporters, "It feels great!" He went on to say, "Before I could hardly raise my head and I could not move. I had to be helped. Now I get up. I walk up on my own. I can help myself. That makes a big difference." Asked about his decision-making process of agreeing to this experimental procedure, Tools replied: "There was no decision to make. I could either sit at home and die or come here and take a chance." One of the biggest adjustments for Tools was getting used to "not having a heartbeat." He had a "whirling sound" from the device, but as he

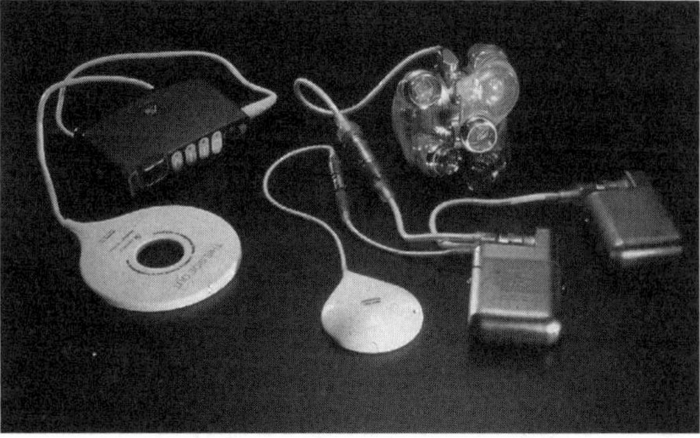

Fig. 1.3 The AbioCor total artificial heart, implanted in a human as a permanent implant in 2001 (Courtesy of the Smithsonian Institution's Division of Science, Medicine & Society).

joked, Tools said the new sound "makes me realize that I am alive." The discourse around the prognosis of the patient was all about returning home, playing with the grandkids, and going fishing.[59] Tools died in November 2001, surviving 151 days with the AbioCor heart.

The second AbioCor heart implant patient, 70-year-old Tom Christerson, became the first artificial heart patient to return home. He was discharged from the hospital within months of the operation, and enjoyed many months in his hometown dining at local restaurants, visiting with friends, and entertaining at home. Christerson died in February 2003, living about 17 months with the artificial heart. The AbioCor heart implant clinical trial is presently ongoing, and many scientists and clinicians (and undoubtedly patients) remain optimistic concerning this next generation of total artificial heart technology.

Decades earlier, the original goal of the artificial heart program was to rehabilitate severely ill heart patients with a reasonable quality of life. Early clinical cases (1969, early 1980s) presented mixed results. Certainly the utility of total artificial hearts as a bridge-to-transplant showed promise by the mid-1980s, and at the same time stimulated greater research on partial artificial hearts or ventricular assist devices for the same purpose. The latest generation of total artificial heart technology and its utility as a permanent implant device have recently presented some spectacular results. To what extent the total artificial heart has come to fruition as a viable technological fix in treating end-stage cardiac failure, only time will tell.

Notes

1. This work is part of an ongoing collaborative history project of the American Society of Artificial Internal Organs, the National Institutes of Health's National Library of Medicine, and the Smithsonian's National Museum of American History, Behring Center, entitled Project Bionics: Artificial Organs from Development to Clinical Use. This is a documentary history project aimed to preserve and document the history of artificial organ developments. The author alone assumes all responsibility for the views, omissions, and errors expressed in this paper.

2. Brief summary of modern surgical developments taken from Frederick F. Cartwright, *The Development of Modern Surgery* (London: Arthur Barker, 1967); Richard Meade, *An Introduction to the History of General Surgery* (Philadelphia: W. B. Saunders, 1968); Robert G. Richardson, *Surgery: Old and New Frontiers* (New York: Charles Scribner's Sons, 1968); Owen H. Wangensteen and Sarah D. Wangensteen, *The Rise of Surgery: From Empiric Craft to Scientific Discipline* (Minneapolis: University of Minneapolis Press, 1978); W. F. Bynum and Roy Porter, eds., *Companion Encyclopedia of the History of Medicine* (London and New York: Routledge, 1997); Roy Porter, *The Greatest Benefit to Mankind* (London: W.W. Norton, 1997), chap.19.

3. See Renee C. Fox and Judith P. Swazey, *Spare Parts: Organ Replacement in American Society* (New York: Oxford University Press, 1992); Renee C. Fox and Judith P. Swazey, *The Courage to Fail: A Social View of Organ Transplants and Dialysis* (Chicago: University of Chicago Press, 1974); Wilfred Lynch, *Implants* (New York: Van Nostrand, 1982); Janice Cauwels, *The Body Shop: Bionic Revolutions in Medicine* (St. Louis: C. V. Mosby, 1986); Katherine Ott, David Serlin and Stephen Mihm, eds., *Artificial Parts, Practical Lives: Modern Histories of Prosthetics* (New York: New York University Press, 2002).

4. American Heart Association, *2002 Heart and Stroke Statistical Update* (Dallas, TX: American Heart Association, 2001), p. 4.

5. J. B. O'Connell and M. R. Birstow, "Economic Impact of Heart Failure in the United States: Time for a Different Approach," *Journal of Heart and Lung Transplantation* Vol.13 (1994): S107–S112.

6. A. Carrel and C. Lindbergh, "The Culture of Whole Organs," *Science* Vol. 81 (1935): 621.

7. C. Hallowell, C., "Charles Lindbergh's Artificial Heart," *American Heritage of Invention and Technology*, Vol. 1 (1985), pp. 58–62.

8. A. Carrel and C. Lindbergh, *The Culture of Organs* (New York: P. B. Hoeber, 1938).

9. *Time*, Vol. 24 (13 June 1938).

10. J. M. Fenster, "The Inner Limits," *Invention and Technology* (winter 1998): 12. George Pantalos stated that "The alliance between Lindbergh and Carrel is of particular note since it represents the advent of the engineer into the clinical domain." Taken from George M. Pantalos, "A Selective History of Mechanical Circulatory Support," in T. Lewisand T. R. Graham, *Mechanical Circulatory Support* (London: Edward Arnold, 1995) p. 4.

11. See A. Weisse, "Turning Bad Luck into Good: The Alchemy of Willem Kolff, the First Successful Artificial Kidney, and the Artificial Heart," in *Medical Odysseys* (New Brunswick, NJ: Rutgers University Press, 1991) pp. 87–105.

12. T. Akutsu and W. J. Kolff, "Permanent Substitutes for Valves and Hearts," *ASAIO Transactions* Vol. 4 (1958): 230.

13. See W. J. Kolff, *Artificial Organs* (New York: Wiley, 1976); Willem J. Kolff, "Artificial Organs beyond the First 40 Years," *Life Support Systems* Vol. 2 (1984): 1–14.

14. T. Akutsu and W. J. Kolff, "Permanent Substitutes for Valves and Hearts," *ASAIO Transactions* Vol. 4 (1958): 230.

15. John T. Watson, "NHLBI Program History," in *Report of the Workshop on the Artificial Heart: Planning for Evolving Technologies* (NHLBI, 1994), pp. 27–32.

16. U.S. Congress, House Subcommittee on Appropriations, *Hearings on Department of Health, Education, and Welfare Appropriations for 1966* (Washington, DC: Government Printing Office, 1965).

17. J. R. Hogness, M. VanAntwerp, eds. *The Artificial Heart: Prototypes, Policies, and Patients* (Washington, DC: National Academy Press, 1991), pp. 205–206.

18. Thomas Preston, "The Artificial Heart," in *Worse Than the Disease: Pitfalls of Medical Progress* by Diana B. Dutton (Cambridge: Cambridge University Press, 1988), pp. 97–99.

19. J. R. Hogness, M. VanAntwerp, eds. *The Artificial Heart: Prototypes, Policies, and Patients* (Washington, DC: National Academy Press, 1991), Executive Summary, pp.1–13 and chapter 1, pp.14–25.

20. I thank J. Ritchie Garrison, University of Delaware for this point as well as other suggestions on this paper in his capacity as panel commentator, "Objects as Fixes" Panel, The Technological Fix Conference, Hagley Museum and Library, 4 October 2002.

21. Thomas Preston, "The Artificial Heart," in *Worse Than the Disease: Pitfalls of Medical Progress* by Diana B. Dutton (Cambridge: Cambridge University Press, 1988), pp. 94–95.

22. Again I thank J. Ritchie Garrison for this comment.

23. American Heart Association, *Your Heart and How It Works*, (AHA, 2002).

24. C. N. Barnard, "A Human Cardiac Transplant: An Interim Report of a Successful Operation Performed at Groote Schuur Hospital, Capetown," *South African Medical Journal* Vol. 41 (1967): 1271; C. N. Barnard, "Human Cardiac Transplantation," *American Journal of Cardiology* Vol. 22 (1968): 584.

25. For more on surgical alternatives to medical therapy, see Peer M. Portner, "Permanent Mechanical Circulatory Assistance," in W. A. Baumgartner, et al., eds. *Heart and Lung Transplantation, Second Edition* (Philadelphia: W. B. Saunders Co, 2002): specifically 531–2.

26. An exception to this is Dr. Michael DeBakey's experimental biventricular bypass pump or functional replacement pump. This device involved two pneumatically-driven silastic sac-type pumps implanted for biventricular bypass. The diseased heart remained intact, and the two blood pumps were implanted in a right-and-left heart bypass placement to maintain blood circulation. See M.E. DeBakey et al., "Orthotopic cardiac prosthesis: preliminary experiments in animals with biventricular artificial hearts," *Cardiovascular Research Center Bulletin* 1969 Apr–Jun; 8(4): 127–42.

27. John T. Watson, "NHLBI Program History," in *Report of the Workshop on the Artificial Heart: Planning for Evolving Technologies* (NHLBI, 1994): 28.

28. Again, I thank J. Ritchie Garrison for this comment.

29. BBC interview transcript, cited in *Scientific American Frontiers: Affairs of the Heart* (PBS-NOVA production, 1999).

30. D. A. Cooley, D. Liotta, G. L. Hallman, R. D. Bloodwell, R. D. Leachman, J. D. Milam, "Orthotopic Cardiac Prosthetic for Two-Staged Cardiac Replacement," *American Journal of Cardiology* Vol. 24 (1969): 723–730.

31. M. E. DeBakey, "The Odyssey of the Artificial Heart," *Artificial Organs* Vol. 24 (2000): 405.

32. National Heart and Lung Institute, *Cardiac Replacement: Medical, Ethical, Psychological and Economic Implications* (Bethesda, MD: DHEW Publ. No. [NIH] 77-1240, October 1969).

33. Thomas Thompson, "The Doctors' Bitter Feud," *Life* (10 April 1970): 72.

34. National Heart Institute, *Artificial Heart Program Conference Proceedings*, Washington, DC, 9–13 June 1969. See also National Heart Institute, Ad Hoc Task Force on Cardiac Replacement, *Cardiac Replacement: Medical, Ethical, Psychological and Economic Implications*, October 1969.

35. Half-way versus full technology implications first espoused by Lewis Thomas and further applied to the artificial heart research program in L. Thomas, *Aspects of Biomedical Science Policy. IOM Occasional Paper* (Washington, DC: Institute of Medicine, 1972); See also Hogness, VanAntwerp, eds. *The Artificial Heart*, pp. 57–58

36. Hogness, VanAntwerp, eds. *The Artificial Heart*, p. 58.

37. Redirection was limited in the sense that the artificial heart program was not abandoned, nor broadened to debate nonmedical issues. Barton Bernstein argues that dominance by biomedical experts (as well as other factors such as great technological optimism, closed decision-making, and minimal scrutiny by Congress) directed the "misguided quest" for an artificial heart. He compares the 1969 program review (prepared by a predominantly biomedical committee) with the 1972 program review (prepared by a predominantly nonmedical committee), which in the case of the latter report raised the greater ethical, economic, and personal issues surrounding this high technology. See Barton J. Bernstein, "The Misguided Quest for the Artificial Heart," *Technology Review* (November-December 1984): 13–19, 62–63; "The Pursuit of the Artificial Heart," *Medical Heritage* Vol. 2 (1986): 80–100.

38. D. A. Cooley, T. Akutsu, J. C. Norman, M. A. Serrato, O. H. Frazier, "Total Artificial Heart in Two-Staged Cardiac Transplantation," *Bulletin of the Texas Heart Institute* Vol. 8 (1981): 305–319; D. A. Cooley, "Staged Cardiac Transplantation: Report of Three Cases," *Journal of Heart Transplantation* Vol. 1 (1982): 145.

39. See "The Artificial Heart Is Here," *Life* (September 1981): 29.

40. University of Utah Special Collections, The Barney Clark Collection MS 670 Box 1, Folder 30, Correspondence Una Loy Clark to Robert H. Ruby, M.D., 12 Mar 1987.

41. W. C. DeVries, J. L. Anderson, L. D. Joyce, F. L. Anderson, E. H. Hammond, R. K. Jarvik, W. J. Kolff, "Clinical Use of the Total Artificial Heart," *New England Journal of Medicine* Vol. 310 (1984): 273–278.; W. C. DeVries, "The Permanent Artificial Heart: Four Case Reports," *Journal of the American Medical Association* Vol. 259 (1988): 849.

42. For examples of this, see "The Brave Man with the Plastic Heart," *Life* (February 1983): 24–30; Harry Schwartz, "Toward the Conquest of Heart Disease," *New York Times Magazine* (27 March 1983): 42–59; "Feeling Much Better, Thank You," *Time* (14 March 1983): 58.

43. For examples of this, see "Death of a Gallant Pioneer: Barney Clark: 1921–1983," *Time* (4 April 1983): 46; "The End of a Long Ordeal," *Newsweek* (4 April 1983): 83–84.

44. See Margery W. Shaw, ed. *After Barney Clark: Reflections on the Utah Artificial Heart Program* (Austin: University of Texas Press, 1984).

45. For an account from the patient's and family's viewpoint, see The Schroeder Family with Martha Barnette, *The Bill Schroeder Story* (New York: William Morrow and Company, Inc., 1987).

46. Kristen E. Johnson, et al., "Registry Report — Use of Total Artificial Hearts: Summary of World Experience, 1969–1991," *ASAIO Journal* Vol. 38 (1992): M486–M492.

47. J. G. Copeland, M. M. Levinson, R. Smith, et al., "The Total Artificial Heart as a Bridge to Transplantation: A Report of Two Cases," *Journal of the American Medical Association* Vol. 256 (1986): 2991–2996; Kristen E. Johnson, et al., "Registry Report — Use of Total Artificial Hearts: Summary of World Experience, 1969–1991," *ASAIO Journal* Vol. 38 (1992): M486–M492.

48. Kristen E. Johnson, et al., "Registry Report—Use of Total Artificial Hearts: Summary of World Experience, 1969–1991," *ASAIO Journal* Vol. 38 (1992): M486–M492.

49. "The Dracula of Medical Technology," *New York Times* (16 May 1988). See also a similar op-ed piece by Daniel Callahan, "The Artificial Heart: Bleeding Us Dry," *New York Times* (17 September 1988).

50. For a review of the field of mechanical circulatory support systems, see Peer M. Portner, "Permanent Mechanical Circulatory Assistance," in W. A. Baumgartner, et al., eds. *Heart and Lung Transplantation, Second Edition* (Philadelphia: W. B. Saunders Co, 2002): 531–557.

51. Eric A. Rose, et al., "Long Term Use of Left Ventricular Assist Device for End-Stage Heart Failure," *New England Journal of Medicine* Vol. 345, No. 20 (15 November 2001): 1435–1443.

52. Those challenging the uncritical optimism of artificial heart technology include Barton J. Bernstein, "The Artificial Heart Program," *Center Magazine* Vol. 14, no. 3 (1981): 22–41; Thomas A. Preston, "Who Benefits from the Artificial Heart?" *The Hastings Center Report* (February 1985): 7; Arthur Caplan, "The Artificial Heart," *The Hastings Center Report* (February 1982): 23.

53. University of Utah Special Collections, The Willem J. Kolff Collection MS 654, Box 353, File 7, Correspondence W. Kolff to Stephen K. Clark, M.D., 19 December 1983.

54. Expert Panel Review of the NHLBI Total Artificial Heart (TAH) Program, June 1998–November 1999.

55. See Stanley J. Reiser, "The Machines as Means and End: The Clinical Introduction of the Artificial Heart," in *After Barney Clark*, Margery W. Shaw, ed., pp. 168–175.

56. Expert Panel Review, June 1998–November 1999.

57. Barton J. Bernstein, "The Misguided Quest for the Artificial Heart," *Technology Review* (November–December 1984): 13–19, 62–63.

58. Again, I thank J. Ritchie Garrison for these comments.

59. 21 August 2001 News Conference, Jewish Hospital, Louisville, KY; Andrew Wolfson, "'It feels great!' Tools tells reporters," *Louisville Courier-Journal*, 22 August 2001; Dick Kaukas, "An Interview with Robert Tools: Patient Grateful for Artificial Heart," *Louisville Courier-Journal*, 22 August 2001; Darla Carter, "Heart Patient Takes Field Trips, Tries Fast Food," *Louisville Courier-Journal*, 26 September 2001.

2

Plugging in to Modernity: Wilshire's I-ON-A-CO and the Psychic Fix

CAROLYN THOMAS DE LA PEÑA

> Who could foresee that 'I-on-a-co' would so fire the imagination of the medieval-minded multitudes that they would cease celebrating, slip their necks into a noose, turn on an ordinary light-switch, and claim to be cured of every known ailment from cancer to catarrh? [1]

> — Dr. Annie Lyle, 1926

At one time or another, most of us have tried an electric fix. It may have been during an exploratory tour of Brookstone when we sat down on the vibrating recliner to relieve an aching back or tried the battery-powered jet massager to soothe tired shopping feet. Or perhaps we've pursued something seen on an early morning infomercial that promised to shape our abs with six simple electric sensors in a few minutes a day. Even if we've avoided such experiences, we all probably have memories of snuggling up under a warm electric blanket to chase away the night chill or applying an electric heating pad to comfort aches and pains.

Few of us, however, would refer to these experiences as true *fixes*. After all, it is not as if we believe that placing our bodies in contact with mild electric currents produces dramatic physical repair. We may enjoy the heat produced by blankets and pads or the vibrations emitted by chairs or whirlpools, but most of us are aware that the fix, in these instances, results from the product of electricity (the heat or the vibration), not in the electricity itself.

This was not always the case. Early twentieth-century Americans frequently purchased items designed to cure the body through direct exposure to electricity. Electric products such as belts, ankle cuffs, "magnetic shields," and even blankets were sold expressly as therapeutic devices. Promotional materials reveal that the reasoning used to sell such products was expressly electric: physical ailments from sciatica to impotence to arthritis and even hair loss could be cured if one applied a mild current directly to the body over an extended period of time.

Scholars have, in the last ten years, explored the geographies of these devices within studies of American electrotherapeutics .[2] We now know that the amount of electricity generated by these products, typically a low-level galvanic charge, was not enough to cure most ailments, though it may have been enough to relieve one's experience of pain. We also know that it was not merely irregular or unlicensed medical practitioners who experimented with electric curative strategies. For the licensed and unlicensed, the office and street-corner based, electric batteries were a powerful marketing tool in the open marketplace of American medicine. Owning a battery allowed a doctor to treat a patient in a manner that, if not always pleasurable, was at least not painful. The chance to be made well, without the pain often inflicted through "heroic" methods, may have kept many sufferers returning for more. In addition to work done on the technology of devices and methodologies of promotion, we also have accessed records that reveal the eagerness of contemporary consumers. Over a forty-year period, hundreds of thousands of Americans consumed electrotherapeutic treatments, in spite of the best efforts of the American Medical Association's legal team to dissuade them.[3]

What we have not explored are geographies more psychic than physical. If products that generated electricity and claimed to cure did not, in fact, cure in a traditional sense, then why were they purchased? It was not simply the forceful efforts of skilled con artists in manufacturing and production, nor was it the technological ignorance of consumers, though these were certainly factors. A more complete answer emerges once we realize that promotion and education created a climate for consumption; they did not drive the purchases themselves. The reasons individuals purchased these products, strapped them on, and plugged them in are more complicated than a simple narrative of deception and naiveté would reveal. Many of those who designed, promoted, and consumed these products sought psychic fixes that were at least equally important to the physical ones that may or may not have resulted. This essay will explore these fixes through a case study of one electric device, the I-ON-A-CO electric belt.

As an object, the I-ON-A-CO presents an interesting challenge to traditional trajectories of irregular electrotherapeutics. The vast majority of products that could be purchased from unlicensed physicians, including electric belts and *oxy* products, had been dismissed by the 1920s as fakes. One would assume from the American Medical Association's extensive records of its successful campaigns to shut down irregular manufacturers, that it would have succeeded by 1920. By then the AMA had employed for years a professional prosecutor who investigated numerous irregular electrotherapists.[4] By collecting data on false claims made by product promoters, Cramp was able to bring individual lawsuits against the manufacturers for postal fraud. He was largely successful; by 1920 few unlicensed electrotherapists could use the mail to sell their wares, and consumers were increasingly distrustful of unsubstantiated claims.

In spite of such a climate, the I-ON-A-CO was a phenomenally successful product during the two years it was on the market. Conservative estimates taken from company sales records suggest that over 50,000 devices were sold between late 1925 and mid-1927.[5] Additionally, when one accounts for the people who did not purchase I-ON-A-COs outright but instead purchased treatments in I-ON-A-CO centers, the number of individuals treated by the devices is closer to 200,000.[6] Such numbers are particular intriguing when considered in the climate of what was going on in regular medical practice by the 1920s. Whereas earlier electric products had been sold prior to 1910, a watershed for reform in medical practice and education, the I-ON-A-CO appeared a full fifteen years after this period. When one chose to purchase an I-ON-A-CO, one chose it instead of (or perhaps in addition to), regular medical treatment. Within this context, the I-ON-A-CO emerges as the single electric belt that people selected over more legitimate medical practices.[7] If the I-ON-A-CO was not their only option, and if in purchasing it consumers went against expert medical advice, the reasons behind their choosing to do so are worth investigating.

The material of the belt itself offers few clues to those interested in explaining its success (Figure 2.1). It was roughly 15 inches in diameter, wide enough to fit easily over the shoulders of a grown man or woman. The exterior was covered by a thick layer of leather that came together in seams at the upper and lower ends. In its interior one could find the single technical part: a thick coil of insulated wire that ran the length of its circumference. Out of this coil emerged a thin cord that connected the interior wire to an exterior power source. The only blemish on the leather casing was the small raised bump from which the cord emerged. Critics of the I-ON-A-CO frequently referred to the device as a "horse collar," presumably to cast upon it an antiquated patina.[7] The characterization largely fits the notably low-tech appearance of the product. In contrast to earlier electric belts, which often featured electric-inspired ornamentation such as lightening bolts and metal shields, the I-ON-A-CO was free of embellishments. One might, from a distance, mistake it for a steering wheel.[8] Further, the device lacked an on-and-off switch; one turned it on merely by plugging it in. As a result, the only suggestion of electricity in its visual presence was a small light on the cord that illuminated when powered. Its weight furthers its image as a traditional device; at two pounds it would

Fig. 2.1 The cover of Wilshire's primary promotional brochure. "I-ON-A-CO the Short Road to Health," box 8, folder 11, Wilshire Family Papers. (Courtesy of the UCLA Department of Special Collections.)

not evoke, for those who examined it or undertook a treatment, the sense of the lightness frequently associated with electricity.[9] Nor would a user or explorer easily be able to discern whether the device was working. Unlike previous electric products, the I-ON-A-CO did not generate palpable signs of electricity beyond its illuminated belt; it never vibrated and it only rarely generated heat.[10] Users frequently asked Wilshire and his agents how they could be sure their product was not defective, since they could not tell after plugging it in whether it was working or not.[11]

The power of the I-ON-A-CO's "fix" is not readily apparent to an observer of the device itself. Rather, it must be placed in the original context in which it was understood. Users, sellers, and makers of the device all came to the product with their own desires. These can be characterized, loosely, as centering on reconciliation, reinstatement, and resolution. Whether the problem at hand was ill health, uncertain finances, or failed ambitions, the I-ON-A-CO functioned as a *Zerspiegel* through which believers could see the world as they needed it to be. By venturing, literally, into its center, they discovered cures whose psychic values quieted doubts about its physiological effects .[12]

Fix Number 1: Wilshire and the Virtue of Profit

The I-ON-A-CO was the brainchild of Henry Gaylord Wilshire. Today Wilshire is better known as an entrepreneur in southern California's early twentieth–century real estate market than as a developer of electric products. In the 1920s, however, Wilshire's name was synonymous with the Iona company and its star product, the I-ON-A-CO life belt. There is much about Wilshire's life that should interest students of early-twentieth-century United States culture and politics.[13] Born in 1861 to a bank president father and socialite mother in the emerging urban center of Cincinnati, Wilshire found himself at the center of several competing systems of political and economic development. At the age of twenty he enrolled in Harvard to study economics. The move suggests that in his early years Wilshire sought to establish himself as a classic capitalist, much in the mold of his father, an original investor in Standard Oil, and his cousins, who would become corporation presidents. He left Harvard without a degree in 1882, making the first of several unorthodox decisions for a son of the eastern financial elite. Along with his brother, Wilshire traveled west to Los Angeles, where he dabbled in a series of speculative ventures before meeting with success in advertising and real estate.[14]

The traditional story told about Wilshire focuses largely on these financial successes. Today, southern Californians are familiar with this Wilshire legacy even if they are unaware of his story. Most have driven down Wilshire Boulevard, the main thoroughfare running from Los Angeles to the coast at Santa Monica past the luxury apartments in Westwood and the toney shops of Beverly Hills. This space marks the legacy of Wilshire's speculative vision: in the 1880s he purchased immense tracts of land during the economic depression that followed a boom and bust cycle. He was, in many respects, skilled in gauging the future. Long before Ford's Model T would make automotive travel the norm, Wilshire bought billboards and arranged to have them placed along the region's first highways. Before the professionalization of city planning he founded the town of Fullerton, turning hundreds of acres of unpopulated orange groves into a center for industry and commerce. A half century before the advent of malls, Wilshire planned a Los Angeles shopping district with integrated shopping, multilane thoroughfares, and off-street parking.

Wilshire was as much a challenger to as he was a beneficiary of capitalism, a fact reflected in his nickname of "millionaire socialist." He ran for congress on the socialist ticket twice, unsuccessfully — in 1890 in California and in 1891 in New York. He founded *Wilshire's Magazine* in 1900 as a forum for the views of radicals and Progressives, and kept it afloat, despite its inability to turn a profit, for roughly a decade. Along with his wife, Mary McReynolds Wilshire, he ran a West Coast salon that was known as an intellectual center for social reformers and hosted, among others, Jack London, Rebecca West, George Sterling, and Upton Sinclair. His private correspondence reveals relationships with the British reformers William Morris and Havelock Ellis as well as playwright George Bernard Shaw. Additionally, Wilshire followed with interest campaigns aimed at assisting the working class, becoming, in the 1920s, a supporter of Mussolini and a critic of United States government policies.[15]

Wilshire's decision, at the age of sixty-four, to end his association with real estate and place his financial resources into the development and promotion of the I-ON-A-CO, has much to do with his refusal to recognize binaries. For much of his life, Wilshire navigated the crossroads of capitalist privilege and socialist sensibilities. In the device, Wilshire found a way to resolve the tension between these two poles. By styling himself as an inventor of a technological device, Wilshire found a space for himself as a public intellectual that was denied to him as a socialist politician. By promoting an expensive device that promised to relieve nearly all physical ailments, yet could be purchased on credit for $5 a month, Wilshire could simultaneously improve the lot of the masses while profiting personally. The result was a technological fix that, at the end of his life, enabled Wilshire to reconcile his contradictory desires for mass utopia and elite privilege.

Wilshire was interested in health cures before he began developing the I-ON-A-CO in 1925. He seems to have created his own health regimens, which he followed consistently as early as the 1910s. Logan, Wilshire's son, recalls in his biography that his father regularly began the morning with a cold plunge and black coffee, reminiscent of an earlier generation of health reformers like John Harvey Kellogg.[16] In the 1920s, Wilshire began looking for a product that would allow him to turn his interest in health into a profit-making enterprise. He explored a number of opportunities, including inventing a silk-based, radium-infused fabric for blankets and undergarments; distributing a bread called "Ex-cell-o" reported to build health and energy; and designing electric heating pads, before settling on the I-ON-A-CO.[17]

Wilshire was not interested in merely distributing someone else's product, even if that product made him money. He briefly did this with Ex-cell-o, sending samples to area stores and encouraging his friends to buy it, but stopped after a few months. Wilshire's real interest seems to have been in developing unique scientific theories about health. His records throughout the 1920s reveal a budding interest in theories of disease and health. In February 1926, just as Wilshire began developing the I-ON-A-CO, he wrote to Thomas Nelson and Sons publishers requesting a $100 eight-volume set of *Nelson's Loose-Leaf Medicine*. That same month he wrote to the dean of an international correspondence school asking for information on an electrotherapeutics course. Evidence suggests that Wilshire spent the years 1924 to 1926 styling his own system of medical education. In 1924 he published an article in the *LA Express* newspaper sharing what he had learned about the cause of hoof-and-mouth disease. By September of 1926 he felt confident enough in his knowledge of physiology to challenge the *American Illustrated Medical Dictionary*. The definition of *feces*, he argued in a letter to its editor, Dr. W. A. Newman Dorland, was incorrect: it was not residue of food but rather intestinal excretions.[18] By January 1927 he was exchanging letters with other self-taught diagnosticians about the nature of disease and the intricacies of human elimination. This was not merely knowledge Wilshire reserved for other experts. He eagerly shared it with his most intimate relations, often insisting that his unorthodox medical theories could solve intractable conditions. When his wife, Mary, began to suffer from sciatica in 1926, Wilshire became convinced that the cause was infected teeth. He admonished her over the next year to ignore the advice of doctors and listen to him instead: if she would only remove her teeth, he argued, she would be well.[19]

Wilshire's records do not include self-reflection on the reasons for his sudden interest in science and medicine in the 1920s. It is not unreasonable, however, to assume that this desire for expertise emerged in response to his political failures. By the 1920s, Wilshire was discouraged with the state of American society. As he informed a supporter, who, in 1924, wrote asking why he no longer published *Wilshire's Magazine*, Wilshire simply had "no constructive proposition to remedy the present ills of society." What he had, he claimed, was only destructive criticism that "would hardly get subscribers."[20] His runs for office had been unsuccessful, and his radical actions for social change (such as his unauthorized speechmaking in Union Square in 1901, which led to his arrest) had made little impression.[21] Even his rhetorical forum for political expertise, *Wilshire's Magazine*, had failed. Within this context, Wilshire's turn to science, and to technological products over which he could assert expertise, makes sense. As sole inventor, developer, and promoter of the I-ON-A-CO, Wilshire found a way to become the activist he was unable to be during his political career.

It is not clear exactly when or why Wilshire came to believe in the particular science behind the I-ON-A-CO. What is clear is that he was exposed to the ideas of German biologist Otto Warburg in the early 1920s and used his own interpretation of Warburg's work to develop a theory of health via magnetized iron. Warburg's work, as described in Wilshire's promotional material, proved that iron acted as a catalyzer that united oxygen to the cells and created oxidation. Warburg's work focused only on the role of iron in the blood, with no specific reference to electricity or magnetism. Wilshire, however, used Warburg's discoveries as the foundation upon which he built his own theory of electromagnetic health. According to Wilshire's main promotional booklet for the I-ON-A-CO, "The I-ON-A-CO, a Short Road to Health," the device worked because it created an electromagnetic field around the body, thereby creating a "catalyzer" that improved metabolism. Within the ring, the body's iron purportedly would be compelled to greater action, speeding up the catalyzation process and increasing the body's absorption of healthy oxygen. Absent the increased magnetic effect, which only the I-ON-A-CO could produce, the body suffered from what Wilshire called "auto toxemia," a disease caused by poisons trapped in the tissues.[22] For the healthy, the I-ON-A-CO helped them better utilize oxygen to improve respiration and remove waste. For the unhealthy, it restored for them the normal process of oxidation.[23]

Wilshire did not rely upon Warburg's theories to promote his device. In fact, he really never used Warburg's theories at all. The biologist's claims dealt specifically with the effects of iron on cell oxidation. Wilshire's claims involved electromagnetism, a phenomenon Warburg did not mention at all, and one related far more closely to the nineteenth-century pseudo-science of Mesmerism than to twentieth-century biology.[24] What Wilshire did do was position Warburg's ideas closely to his own, only not so close as to make Warburg the expert. In an excerpt from the "Short Road" taken from Warburg's article in *Science*, Wilshire quoted him as saying "every living cell contains iron and that life without iron is impossible."[25] How exactly this iron relates to the I-ON-A-CO, Wilshire does not say. Rather than functioning as the basis for the I-ON-A-CO's medicinal effect, Warburg's theories appear on a parallel plane. Readers are left to make the connections for themselves, perhaps through a relationship between the iron and the device's wire coils or through the sound of the name itself. As a result, Wilshire was able to interpret science for an audience unlikely to read *Science* for themselves. In his radio shows, his in-store lectures, and his public presentations, Wilshire presented himself as an expert in practical science.[26] Few of his listeners, especially in small towns where his radio shows often played, were likely to find where his interpretations were incorrect. This was an image he cultivated through photographs as well: in pamphlets and advertisements his portrait appears as a cross between Sigmund Freud and George Bernard Shaw, a man in whom psychology, drama and science converged (Figure 2.2).[27]

Part of the I-ON-A-CO fix was, for Wilshire, the opportunity to be a public intellectual. As he wrote to his friend, Major Haldane Macfall, in 1926, Wilshire saw his struggle for the I-ON-A-CO as closely resembling his political struggles. "I would say that I think almost the same condition prevails in getting it before the attention of the educated, cultured public, particularly the Doctors," he stated, "as I had trying to get Syndicalism before the educated and powerful politicians."[28] His reference to the political system he supported, one based on reforming the means of production, suggests that Wilshire saw the physicians who rejected his I-ON-A-CO as akin to those earlier politicians. In both arenas he found himself barred. Within the medical marketplace, where technological marvels seemed possible and medical professionals were often impotent to police professional boundaries, Wilshire at last found space for his voice.

Part of Wilshire's fix was the attraction of this expertise. The other part was the attraction of virtuous profit. At the same time Wilshire began developing his I-ON-A-CO, he reported to friends and acquaintances that he found himself in difficult financial straits. By the 1920s Wilshire found himself paying escalating expenses for his family. His wife Mary had embarked on an extended tour of Europe; his son Logan had enrolled in

Fig. 2.2 An advertisement for the I-ON-A-CO featuring the "plugged in" woman and the Wilshire cityscape. "New Road to Good Health," box 11, folder 19, Wilshire Family Papers. (Courtesy of the UCLA Department of Special Collections.)

college in England; Wilshire himself maintained his family home in Pasadena and a coastal cottage in southern California. That, combined with a slumping southern California real estate market (he characterized it in April 1924 as "very dull"), provided ample reason to pursue a money-making scheme. Wilshire's preoccupation with money is clear in his personal correspondence. "My family wants money," he wrote to a friend in January 1924, following this with another letter the following day declining an offer to buy into oil speculation because he had "no money for anything." By July of that year the situation appears to have worsened. "We are experiencing a very decided break in our previous prosperity," Wilshire reported, forcasting that "it does not look … as if there will be a change for the better for quite a while."[29]

In fact, evidence suggests that Wilshire's interest in electrotherapy stemmed from this difficult financial position. Along with his friend, novelist Upton Sinclair, Wilshire began research into an invention of Dr. Albert Abrams, in March of 1924, just when his finances were looking their most dismal. Abrams was notorious as the inventor of the Autoclast, a large machine that he claimed cured disease by amplifying specific vibrations in the blood. Abrams encouraged sufferers to take a sample of their blood and send it to him; he placed the blood in the machine, looked at the reading as to which vibrations were off, and prescribed a treatment that was also given by the Autoclast.[30] His unorthodox medical theories, combined with his status as a licensed physician, gave Abrams a good deal of negative publicity, among physicians and laypeople before his death in 1923.[31] Wilshire appears to have considered picking up where Abrams left off, going so far as to writing to friends to generate interest in the machine and asking Westinghouse Electric and Manufacturing Company to make him an Autoclast of his own. Unlike Abrams, and suggestive of the approach he would take with the I-ON-A-CO, Wilshire appears to have considered selling the devices to both physicians and what he called "the laity." The result would be a lucrative exchange: Wilshire planned on paying $50 for the Autoclast's manufacture and selling it for $150.[32]

Money was a factor in Wilshire's absorption in tech fixes. It was not, however, the only factor. For Wilshire, the I-ON-A-CO allowed him to reconcile two disparate elements in his life: his attempts at socialist reform and his attempts at capitalist success. His private evaluations of Abrams' product reveal his own doubts about its efficacy: one letter to a friend laments that Abrams was a "difficult man to submit to a scientific test."[33] One can assume that one of the reasons he did not take the device into production was that he was never convinced it actually worked. His turn to the I-ON-A-CO may, then, have been because he believed that it delivered on the Autoclast's potential: it made money and delivered a universal panacea.

It is possible to use Wilshire's own correspondence to argue that he knowingly promoted a fraudulent product. After receiving a copy of Wilshire's "Iona" promotion booklet in 1925, Allan McIntyre, a reporter whom Wilshire knew well, wrote congratulating him because "every dam [sic] boob in the country will want one of these" after he hears of it. Wilshire, apparently taking pleasure in the comment, forwarded the letter to his friend, the writer George Sterling, stating that he hoped "you will get as much of a laugh out of my circular." Yet Sterling's response suggests that both he and Wilshire were I-ON-A-CO believers. "Profuse thanks for the invaluable information supplied by your circular," replied Sterling, who noted "with grief the skeptical attitude of your misguided friend, McIntyre … he will prove forever opaque to the truth."[34] Wilshire's own insistence that his friends use and endorse his product affirms this more generous reading. At his request, George Sterling and Upton Sinclair risked their reputations by endorsing the device.[35] When family and friends took ill, even with a disease as serious as breast cancer, Wilshire urged them to plug into their I-ON-A-CO cure. His own death in September 1927 appears to have been hastened by his insistence that he could treat what he perceived as arthritis (it was actually heart disease) with the device.[36]

The I-ON-A-CO, by costing roughly $3.50 to manufacture and selling for $65, offered plenty of opportunity for profit. At the same time it also seems to have been intended to deliver on some of the promises that had failed to materialize through Wilshire's political activism. It was a product that was accessible, if not to all, then to most people. He regularly allowed people to purchase the device on credit, allowing those with as little as $5 to take the device home and treat themselves. It allowed people to bypass the control of elite physicians over their bodies by facilitating home cures. And it reportedly physically restored people, returning them to productive power from disabled states. Whether or not the I-ON-A-CO did these things is another issue. In understanding its fix for Wilshire, it is important only to realize that to him it seemed it could.[37] The power of this reconciliation may have been why Wilshire reached for the I-ON-A-CO at the end of his life, despite the well-publicized doubts about its efficacy at that time. Profit, public expertise, and political empowerment may have been interwoven in its coil.

Fix Number 2: Sellers and the Powerful Fix

Wilshire's belt empire was largely a grass-roots organization. It relied on the efforts of hundreds of local sellers across the country. What emerged in 1926 as a local business centered in Los Angeles, had, by mid-1927, expanded into a complex system of distribution that stretched across the United States. There are no definitive figures on how many individuals Wilshire employed as salespeople; records suggest that the numbers were well into the hundreds. Belts were sold in two ways, involving two types of agents. The first were distributors, individuals who were hired to cover particular regions and promote the product through local demonstrations and door-to-door sales. The second were in-house sales associates, usually called *demonstrators*, who worked in I-ON-A-CO regional offices providing free and paid treatments in addition to selling belts. The first group was the largest. Business records reveal that Wilshire allowed forty-one individuals to represent the belt in southern California alone in 1926.

Used as a model, such a figure suggests that there were hundreds of individuals actively representing the product in cities where Wilshire allowed distribution, including, by 1925, Everett, Washington; Palm Beach, Florida; Covington, Kentucky; Wheeling, West Virginia; Pomona, California; and Cincinnati, Ohio.[38] The second group, while smaller in number, had the most contact with Wilshire himself and with prospective buyers. Each of Wilshire's demonstration offices employed two or three individuals whose job included greeting prospective customers, applying I-ON-A-COs for treatments, and selling devices for home use. By 1927 Wilshire had twenty-three regional offices, making the number of demonstrators employed probably close to seventy.[39]

While Wilshire's sellers differed in region and sales approach, they shared common desires. It is impossible to say, precisely, why men and women joined the I-ON-A-CO venture, and few records survive that would reveal the specifics of the choice and reflections that individuals had upon their jobs as agents and demonstrators. The records that do survive, however, allow us to theorize that sellers were also in search of a technological fix. By connecting their bodies and reputations to the device, these individuals sought to overcome barriers to success based on class, gender, and economic standing.

Both agents and demonstrators seem to have joined forces with Wilshire for economic gain. For individuals like Clara Nave, an old acquaintance of Wilshire's, the device seemed in 1926 a good investment as she found that there was "on all sides, great enthusiasm" over it. For G.A. Sarles of Portland, who wanted local distribution rights, Wilshire's "scientific explanation" and testimonials convinced him that it "should be a very salable device." It is logical that investors and distributors would have been focused on future economic gains; both groups had to make cash outlays without guaranteed returns on investments.[40] This concern, however, motivated in-house demonstrators as well. A Mrs. Hall of Fresno, California, wrote to Wilshire in 1926 after seeing the device in a local office. She and her husband had taken some treatments, but her real interest was in being hired as a demonstrator where there was a chance to earn money. Her financial stresses weighed on her sufficiently that she appealed to Wilshire by detailing their private difficulties in hopes of landing the job: they really needed the work, she claimed, since "we have had so many financial losses."[41]

These accounts of economic troubles reveal more than financial motivations. Each also suggests a desire to engage with a segment of society from which they were excluded. Nave mentions the enthusiasm around her, suggesting that her involvement with the belt might make her part of that same excitement; Sarles talks about testimonials from satisfied users and "scientific" evidence, suggesting that selling the belt would be a way of identifying with its expert backers; Hall contrasts the effective device she experienced with her financial losses, suggesting that one environment might offset the other. Each, in a different way, can be understood as connecting with the I-ON-A-CO in order to gain status. Had they desired only to make money, the letters would not have had the flavor of testimonials. There is something of a religious fervor in the ways people expressed their desire to promote I-ON-A-COs. Arguably, part of the attraction in connecting one's fortune to the belt was the chance to become a more integral part of modern life, one that was changing dramatically in the 1920s.[42]

We can read this desire more fully by exploring the reasoning used by others who, like Nave, Sarles, and Hall, did not share in the era's economic boom. For Adelaide Fenton of Coronado, California, the I-ON-A-CO seemed a good way to combat the effects of a recent divorce. Her request to sell the product contains the blunt admission that she's "not going to get much" from her husband in a pending divorce.[43] Her stationary, with personal letterhead, combined with her residence on the upscale island, suggests that Fenton has enjoyed a position of social status. Read against this background, her request to sell the I-ON-A-CO can be seen as an attempt to take control of a situation over which she had little social or political power. Selling the I-ON-A-CO, perhaps to the very same people she leisurely socialized with during her marriage, was a way to maintain her standard of living. It may even have allowed her to trade one status for another, higher, one: rather than being someone's wife she was someone's healer. A similar desire for renewed status drove Adolph Linsebarth who, like Fenton, wrote to Wilshire in January 1926. His request to distribute the device included a brief biography that described him as a sixty-six-year-old foreign language editor who has not "seen much work since the war."[44]

For Linsebarth and Fenton, the I-ON-A-CO presented an opportunity to reengage with a society that seemed to have left them behind. Linsebarth's statement hints that he once performed skilled and valued work. His allusion to the war connotes valor and service; his credentials as an editor suggest extensive education. Yet as a sixty-six-year-old unemployed man, Linsebarth's sense of his value had lost its foundation. Like Fenton, the conditions of success had changed and he was not doing well under the new rules. For individuals in these positions, investing and distributing the device may have allowed them to regain a position of importance. Fenton could heal bodies rather than play hostess; Linsebarth could promote a technological curative rather than being regarded as a man whose best years had passed.

The chance to be valued is most dramatically illustrated by the experience of Mrs. B. Bredemus, a demonstrator in the Palo Alto office in 1926. In a letter to Wilshire documenting what she judged to be unfair treatment from her supervisor, Mrs. Bredemus provided a sense of the tenuous status staff members enjoyed through association with the device. Bredemus identifies herself as a "nurse" and describes her job as testing and maintaining the "machines," greeting customers and applying the machines for treatment. Wilshire's demonstrators regularly referred to themselves as "nurses" (most were female), though their medical training consisted only of what Wilshire had taught them or provided them in the form of I-ON-A-CO promotional brochures.[45] Her letter suggests that she identifies as working class: she comments on the superior standing of clients who come into the Palo Alto office and claims that she "knows her place" and how to make them feel important.[46] Yet in spite of her admitted "inferior" position, Bredemus feels able to write to Wilshire directly and dispute her supervisor's assessment of her value. "People have said to me if it were not for my personality and pleasant manner, they would not come in," she explained, refuting claims made by her supervisor that she is too slow and hurting business. She comments on her efficiency in treating clients and maintaining the machines, and claims that often, for hours a day, she runs the center by herself. "I am there some mornings till 10:30," she writes, stressing that without her the businesses "wouldn't have grown."

The I-ON-A-CO allowed Mrs. Bredemus to see herself as more than a custodial agent or replaceable salesperson. In her mind, as her rhetoric attests to, she was a "nurse," one trained in the skilled operation of machines and the treatment of ailing bodies. It is this perceived position that allows her to challenge those who have authority over her: her supervisor, and by association, Wilshire himself. The power of position bestowed upon those who believed by other believers should not be overlooked in a rush to judge the I-ON-A-CO as entirely fraudulent. There was certainly real power in dressing for work each morning in a nurse's uniform, in presenting oneself as a therapeutic expert, in operating a machine for a living when few women did. Something beyond the need for a regular paycheck motivated Bredemus to sit and write a ten-page letter documenting her service to the office; certainly she could have found another job in sales had she desired. Her refusal to be dismissed, combined with her evocation of her own professional status, suggest that the device was an important fix for people like her in offices across the country. Power and position denied in regular medicine and in regular employment could be enjoyed as an I-ON-A-CO professional.[47]

Fix Number 3: Consumers "Plug In" to Modern Life

As an object, the I-ON-A-CO is unusual for its time. Unlike previous electric products, its design evokes little sense of the electric fantastic. There are no lightening bolts on its edges, no metallic strips running the length of its leather casing. It leaves a modern viewer to wonder whether this was, in fact, understood as an electric product at all. After all, once the user actually put his or her body in contact with the device, there was no vibration, no heat (typically), and no visual reminders of its electric nature. It appears, in many respects, the "horse collar" it was once labeled: a product decidedly untechnological.

This reading reflects the importance of analyzing the I-ON-A-CO within its original consumer environments. People may have seen the device as a solitary object in advertisements, on the arms of door-to-door salespeople, and even sitting on neighbor's couches. They used the device, however, to connect themselves directly to urban and domestic electrical systems. In fact it may be this very connection that kept the device unadorned. Unlike earlier electric belts that were powered by batteries, the I-ON-A-CO had a plug. As a result, a user interacted with Wilshire's device, perhaps not by staring at it as they would with a battery-powered belt given its galvanic power source, but by staring at the wall from which the current generated. The device was not the power but rather the medium by which power reached the body. This understanding of the I-ON-A-CO is essential if we want to understand how it may have provided a fix for consumers in spite of its questionable medicinal value. Treating your body with the I-ON-A-CO meant plugging in. In an era where few individuals had direct contact with industrial or domestic electrical systems, this might have supplied psychic relief.

Consumers used the I-ON-A-CO in two primary spaces: company treatment centers and private homes. By late 1926 there were at least fourteen major treatment centers; some historians have estimated that the number, nationwide, was closer to twenty-five.[48] The cost of opening each center, estimated at $10,000 by Wilshire in 1926, allows us to theorize that these locations would have been extensively discussed before sites were selected.[49] It is then significant that each center was located in a dense, urban area and each was placed within a building that was associated with the character of downtown. Typical of many were the offices in Los Angeles, Oakland, and Seattle. All were located on predominant streets that could count on foot traffic flowing to and from work. All were located in notable buildings that were considered important downtown architectural spaces.[50] Downtown locations are logical spaces to promote products whose sales relied almost entirely upon newspaper advertisements, word-of-mouth, and foot traffic. Yet the locations were equal parts sales strategy and product positioning. I-ON-A-CO advertisements allow us to theorize that these offices were meant to materially link the product with urbanity.

Wilshire's most successful office was his San Francisco center, located at 150 Powell Street.[51] To attract business to a space, Wilshire placed a series of advertisements in local papers, including the "New Road to Good Health (Figure 2.2). The advertisement contains three elements: the woman receiving a treatment, the text describing the device, and the photo of Wilshire in the top left corner. The dark backdrop makes the photo of Wilshire particularly prominent, as is the white line sketch of Wilshire Boulevard that frames his portrait. Reading the accompanying text, one learns that the device displayed below is directly connected to the urban environment sketched above. Wilshire is the mediating body here, a man described as the "originator of beautiful Wilshire Boulevard, Los Angeles." Were Wilshire merely trying to communicate his financial success, there would be no need for the rendering. Arguably the illustration, complete with towering skyscrapers of the Los Angeles skyline in the background, is designed to present Wilshire *and his device* as agents of modernity.[52]

This was an association that confronted prospective consumers in urban and rural environments. In Long Beach, California, Wilshire launched a billboard campaign placing his face and his belt up and down highways heading into major urban areas. In smaller cities like Lawrence, Kansas, and Boise, Idaho, Wilshire used radio to broadcast his belt into local living rooms. In both urban and rural environments, Wilshire placed an advertising slide in movie theaters where it would show up on screen before the film was shown. Wilshire also accompanied the I-ON-A-CO at expos, such as the food show in Los Angeles, where the product was featured among the latest innovations in packaging, preparing, and engineering.[53]

These sites provided prospective consumers with ample opportunity to associate the device with modern technology and urban life.[54] In the late 1920s, radios were still relatively new technologies designed to broadcast the city into small towns and rural regions. Movies remained novelties that entertained with content and technological delivery. Billboards were only visible to those passing in cars and on highways, objects and spaces that had only existed for a generation. Expos were spaces where people went specifically to see new things and preview the future. It is likely that the "several hundred" who took I-ON-A-CO treatments at the Los Angeles food show understood it as part of the same modern project as the innovative products that framed it. Part of what left them, in Wilshire's words, "completely restored to their original vitality," may have been the experience of seeing modern packaging, modern distribution, and modern "medicine" under the same roof.[55] This was a connection made explicitly in promotional materials, often with the specific aim of countering doubts about Wilshire's ability to have created a cure-all. "Naturally all of this seems beyond belief," one brochure told readers, "but the radio, sending winged words around the world, the airplane, carrying Byrd to the North Pole, would have been beyond belief a few years ago."[56] Wilshire's device, seen in this light, was just another link in the chain of modern inventions by a man who built a metropolis out of the desert and envisioned the highway before Ford. If one could believe the Wright brothers could make thousands of pounds of iron soar, one could believe that Wilshire could cure the body with coil and a power outlet.

In treatment centers the I-ON-A-CO offered a fix that extended beyond a mingling of technology, modernity, and material. It also made bodies the sole beneficiaries of the collaboration. Within treatment rooms, patients were taught that their bodies could be infused with the best that medicine and technology had to offer. This is illustrated by recreating the typical experience a customer would have had at an I-ON-A-CO center. It would have begun, often, when one observed the treatment center from the street, typically visible above a neon sign. After entering the reception area, they were taken by an attendant to one of two public treatment rooms. An advertisement for the San Francisco center shows a photograph of a typical treatment room in the San Francisco office. We can assume that it is artificially crowded with well-dressed patients; Wilshire probably stocked the space in order to make it look as if respectable middle-class men and women were crowding the facility. Nonetheless, the photo illustrates how Wilshire intended his offices to be used, and how they typically were run, given his tendency to drop by and evaluate his offices frequently. Here, individuals line the walls of the room, attended to by two women and facing inward where two men in suits (probably demonstrators) sit next to a bearded man who appears to be Wilshire himself (Figure 2.3). The suited men stand directly next to the I-ON-A-CO, seemingly engaged in a product demonstration as Wilshire looks on, hands in his lap, happily suggesting his inventor status. There are several notable things going on with the patients who line the walls. First, they are sitting in close proximity to each other. Second, several of them gaze downward, apparently reading promotional materials of some sort. Third, three of four of the individuals are looking up as if they are taking instruction from the "nurses" in the room. Fourth, every individual in the space wears an I-ON-A-CO around his or her shoulders or waist, a device that we can assume is plugged in to the wall behind them, beyond our view.

The scene is one where technology, medical science, and urbanity are all placed in service of the average bodies that line the walls. One can read it as a sandwiching of power sources benefiting the consuming bodies. On one side are the forces of medical expertise (in the form of Wilshire's self-appointed "nurses" and the white-collared men who hold the device) and business acumen (in the form of Wilshire himself). On the other side are the wall sockets that link the energy flowing from the city to physical bodies through the connecting I-ON-A-CO device. Each individual is directly mediating, with his or her body, these two forces of modern power. Whether they are reading a brochure that describes the power of magnetizing iron and catalysts, talking with the nurses whose medical educations are limited to what Wilshire has taught them, or listening to Wilshire and his physician experts directly, the patients are being taught that the device they sit under delivers the latest advance in scientific knowledge. And each began her experience in this space by watching as an attendant selected the seat, draped an I-ON-A-CO over her shoulders, and plugged her into the wall.[57]

I-ON-A-CO treatments taught consumers that electricity benefited the body.[58] Notably, the very first thing prospective patients did when they entered was fill out a detailed medical questionnaire. The two-page docu-

Fig. 2.3 An advertisement for the I-ON-A-CO treatment centers featuring the San Francisco office. "Truth, the Basis of Virtue," Wilshire I-ON-A-CO correspondence, p. 2, The Iona Company, folder 0403-13. (Courtesy of the American Medical Association Archives.)

ment asked people to evaluate the healthiness of their bodies and record the results they found. Here, before their treatments, patients considered and reported their height and weight, and whether they had high blood pressure, lung trouble, or diabetes.[59] Only after they provided a detailed description of their physical health did the attendant give the I-ON-A-CO treatment. Arguably, this period of forced physical self-reflection made customers hyper-aware of the physical contact between their depleted bodies and the device's ample electricity.[60] They were not merely coming in, placing a device about their shoulders, and enjoying the novelty of getting plugged in. By the time the plug was inserted, they had ample opportunity to reflect on their own physical imperfections. And they had ample time to build an expectation that plugging in would fix them.

In addition to the fixes provided in I-ON-A-CO offices, customers also were able to navigate domestic technology using the device at home. It is telling that promotional pamphlets, newspaper advertisements, and even letterhead frequently show a variation of the plugged-in woman. Letterhead from the Fresno office from 1926 included one variation with every letter advertising the product or delivering it to consumers (Figure 2.4). "I-ON-A-CO: the Short Road to Health," Wilshire's main promotional brochure, sent to all who inquired about the device and direct mailed to many who did not, also featured a version (Figure 2.1). In each image, a modern woman (signaled by her short haircut and modern dress — sleeveless in the Fresno letterhead) sits undertaking some household task while plugged into the wall behind her via her I-ON-A-CO. The illustrations are important, not merely because they were nearly ubiquitous in product promotions, but because they offer insight into a psychological cure that the

GRAYLORD WILSHIRE, President

PHONE 6366

Wilshire's I-ON-A-CO

THE NEW WONDER CURE

1926 MARIPOSA STREET

FRESNO, CALIFORNIA

December 18th 1926.

Mr. Gaylord Wilshire,
Meuhlbach Hotel,
Kansas City, Mo.,

My, Dear Mr. Wilshire;

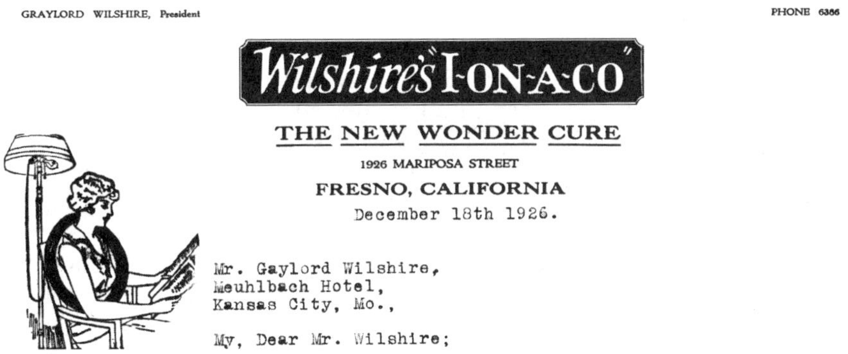

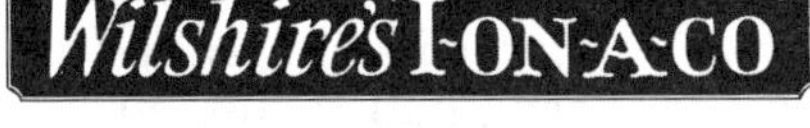

Fig. 2.4 Letterhead from the Fresno I-ON-A-CO office, box 8, folder 4, Wilshire Family Papers. (Courtesy of the UCLA Department of Special Collections.)

device may have offered home users. Assuming that consumers saw these illustrations and associated their own experiences with what they observed, the I-ON-A-CO appeared to be a means to simultaneously render electricity safe and confer expert status upon its users.

It is significant that these illustrated characters are women doing domestic labor. Many I-ON-A-CO purchasers seem to have been female, so the strategy might have been purely to mirror an important consumer base. Yet several features of these illustrations suggest that they have allegorical value beyond purely selling to consumers who want to see a representation of their sex. In fact, the female figure, when considered as an element in a larger scene, may have appealed equally to women and men. One can see the projects undertaken by the woman, reading a book and embroidering a blanket, as symbols of domestication .[61] At the same time, the light behind the female figure also functions as a symbol of domestication, connoting family and hearth.[62] When we combine these advertising images with the actual experience of using the I-ON-A-CO, it becomes possible to argue that some of the attraction in using the device was the experience it afforded to safely engage with electricity firsthand.

When selecting prospective consumers to court, Wilshire often directly targeted people who received residential power service. Before opening offices, local agents often sent introduction letters and promotional literature to people who subscribed to phone service or who were customers of the local electric company.[63] Such a strategy had, of course, a functional purpose: it allowed them to target only those individuals who could actually plug in an I-ON-A-CO at home. Yet it also had a psychic purpose. It delivered, directly to people whose domestic environments had recently been reshaped by electric power, the message that electricity was both controllable and beneficial.

By purchasing and plugging in an I-ON-A-CO, customers were able to experience this physical lesson for themselves. Brochures devoted a good deal of space to discussing the distinctions in electric currents and evaluating the ones on which the device could run (either 110 volt AC or 32 volt DC). Further, they clearly detailed how customers should plug the device into their home power sources. Among the parts included with an I-ON-A-CO, reported the "Short Road to Health," was the cord and attachments necessary for connecting with a variety of outlet types in the 1920s home, including lamps, sockets, walls, and floor plugs. People were instructed to test the device after they plugged it in to ensure that electricity was flowing through the ring. This entailed using a small test ring included in the box which, when held parallel with the I-ON-A-CO, lit up a small bulb indicating the electric flow. Should customers be using AC current, which would not light the bulb, they were encouraged to test for electric presence by placing an object of iron or steel within the circle to magnetize it. Were they to undertake that experiment, the brochure warned, they would need to first remove their watch so as not to magnetize it as well.[64]

These instructions made physical interaction with domestic electricity necessary to receive an I-ON-A-CO cure. Not only did people have to figure out what kind of current they had in their homes, but they also had to actually unscrew light fixtures, find correctly sized adaptors for the device's cord, screw it into the wall, and then wrap it around their bodies. This experience was, for many customers, probably their first attempt at understanding and altering their home electric systems. When Wilshire's brother wrote to him saying that he would try the device, he reported that he had verified the current in his home as 110 volt, 60 cycles, AC. This statement suggests that he had read Wilshire's promotional material discussing the DC and AC models, and had taken the time to consider his own home system and order the correct product model. Leonard Abbott exchanged two letters with Wilshire in 1926 attempting to figure out his electrical system so he could use the I-ON-A-CO. "I have no idea what the voltage and current of the electrical system used in this house is," he commented, asking "How does one find out that sort of thing?" Ten days later he wrote again, saying that he had discerned that there was "direct current here," because he had not been able to ignite the test bulb, something that Wilshire's brochure indicated happened when one's home had direct current.[65]

Users of the I-ON-A-CO were able to decode their electrical systems using Wilshire's promotional pamphlets and direct correspondence with distributors. In an age of limited electrical exposure, this afforded the opportunity to define oneself as an expert, perhaps dispelling a sense of confusion in a rapidly modernizing age.[66] Further, the device, in use, allowed consumers to see electricity as not a dangerous force knowable only by experts. It was, instead, a benign agent of health that was easily manipulated at home (as illustrations revealed: even a woman could do it). Annie Barton, who used a copy of the I-ON-A-CO promoted after Wilshire's death in 1927, felt that the electrical connection was the essential part of treatment worth. "One sits at ease with an appliance (like a small auto tire) about one's waist which is plugged into an electric light socket," she explained to a female friend. Yet there was no need to worry, she continued, as the current was controlled and produced no electric shock.[67] The opportunity for those like Barton, who were excluded from the ranks of electric professionals, to master electricity physically and mentally may have been a powerful psychic fix.

Conclusion: New Hope and Courage in the I-ON-A-CO

An entirely separate essay could evaluate the curative claims made by Wilshire and his promoters for the I-ON-A-CO. This one has not tried to argue that the devices actually cured people. Such a statement is highly unlikely given the low-level current used in the product. In reality the I-ON-A-CO was not so much a "horse collar" as it was a circular heating pad. What I have tried to do is explore the ways in which this product appeared as a fix for a number of discomforts of the age: the discomforts of abandoning one's socialist principles; the discomfort of finding one displaced due to gender, age, and economy; the discomfort of domestic electrification by which an unknown, possibly dangerous force was suddenly brought into the home, a site held nearly sacred as a protected refuge from danger and urbanity since the mid-nineteenth century.[68] This psychic fix may, in the end, have been the only real cure delivered by the device, and the one most ardently believed in by its promoter.

In 1927, months before Wilshire's own death at the age of sixty-seven from heart disease, one of his associates circulated a memo to regional offices urging them to encourage those who had purchased I-ON-A-COs for home to come into the local offices for further diagnoses and demonstrations. The anonymous author speculated that in doing so they would have greater contact with the company's medical department. To justify this extra effort on the part of the local staff, he or she cited "the psychological value of this co-ordination."[69] The mention of this value was not merely an idea promoted by company outliers. Wilshire himself appears to have been aware of this psychological element in the device's cure. Nearly a year before that memo, he had written to his closest physician ally, Dr. David Starr Jordon of Los Angeles, urging him to try the device because "there is no question about the psychological effect whatever the physiologic effect may be. They [patients] get in this frame of mind immediately and stay there for three months without change, in fact getting more and more cheerful every day."[70]

The idea that a frame of mind might influence one's sense of physical health was not an idea that Wilshire came upon haphazardly. Mary McReynolds Wilshire, Wilshire's second wife whom he married in 1904, was one of the first to practice Jungian psychoanalysis in the United States. In 1916 she studied with Jung in Zurich; by 1917 she was described as the method's "chief exponent in this country." Mary seems to have spent 1917 through the early 1920s practicing analysis in private practice, teaching courses in southern California, and attempting to begin a school for analysis in Pasadena. Her work has not yet been analyzed by historians, in spite of the effect she must have had on some of the most well-connected citizens of the region. Her clients included Mrs. Lucy Fiske, Mrs. Henry James, and Mary Putnam.[71]

Reports on Mary's personality leave little doubt that she had a profound impact on many around her, including her husband. She was described in 1917 as "a woman of astonishing magnetism and vivid personality" who had an exceptional ability to "put into simple terms the difficulties and perplexities of human mentality."[72] Her unpublished manuscript, titled "A New Conception of the Human Mind, Its Aims and How It Functions," leaves little doubt that Mary was a keen student of the ability of the mind to influence the body. "It is my life work to bring order out of mental chaos," she explained, "to set feet which were groping in the morass of doubt on the highway of happiness, contentment and success; to give new hope and courage to those from whom the light of proper self expression was shut off."[73] During the same years that Wilshire was investigating the work of Warburg and working with Westinghouse to develop his I-ON-A-CO, Mary was working as an astute student of the psyche, analyzing more than fifty thousand dreams by the mid-1920s by her own account.[74] It is likely that Wilshire was influenced, at least in part, by Mary's "new conception" of the human mind. At the very least, Jungian analysis was a familiar terrain to Wilshire *before* he even conceived of the I-ON-A-CO. For both Wilshires, cures may have been sought by accessing unconscious fears, subjecting them to modern treatments, and allowing the mind to lead the body to health.[75]

For its developer, its sellers, and its consumers, the I-ON-A-CO offered an unorthodox approach for navigating a number of cultural changes embedded in the mid-1920s. So strong was the product's appeal that in spite of its very questionable efficacy and the American Medical Association's efforts to reveal it as fraudulent, versions of the device continued to sell well after 1927 when Wilshire's sudden death lead to the dissolution of the Iona company and, as a result, the demise of the I-ON-A-CO. Sellers like Philip Isley, the manager of the Cleveland office, continued to promote the product, in Isley's case under the new name of Theronoid. As late as 1933, nearly ten years after its appearance, imitations of of the "life buoy" were still finding promoters, sellers, and customers. The device's longevity, as that of many pseudo-medical devices of that era, and of our own, begins to make sense once we view it as a deliverer of cures more psychic than physical. As Mary might put it, having the power to confront your fears might enable you to leave behind "the morass of doubt" and embrace instead "new hope and courage." This may in the end have been the I-ON-A-CO's most legitimate technological fix.

Notes

1. Anna Lyle to Arthur Cramp, December 1, 1926. Correspondence in folder 0403-11, AMAHHFC. Lyle seems to have converted to Wilshire's method. In a quotation that I have been unable to verify, Lyle reportedly later endorsed the I-ON-A-CO in the *Ionaco News*. According to Stuart Holbrook, Lyle wrote that she had "fallen for the I-ON-A-CO strongly" and found it effective in treating David Starr Jordan, president emeritus of Stanford University. Stewart Holbrook and Eric Jameson, *The Golden Age of Quackery* (New York: MacMillan, 1959), 141–142.
2. See, for example, Eric Jameson, *The Natural History of Quackery* (Springfield, IL.: Charles C. Thomas, 1961); Holbrook, *The Golden Age of Quackery*; Margaret Rowbottom and Charles Susskind, *Electricity and Medicine: History of their Interaction* (San Francisco: San Francisco Press, 1984); Lisa Rosner, "The Professional Context of Electrotherapeutics," *Journal of the History of Medicine and Allied Sciences* v. 43, n. 1 (1099): 64–82; and Carolyn Thomas de la Peña, *The Body Electric: How Strange Machines Built the Modern American* (New York University Press, 2003).
3. For more information on other products that were manufactured and purchased at the turn of the century see Thomas de la Peña, *The Body Electric*.

4. These records were published by that investigator, Arthur Cramp, in three volumes. See Arthur Cramp, *Nostrums and Quackery*, vol. 1 (Chicago: American Medical Association Press, 1912; Cramp, *Nostrums and Quackery: Articles on the Nostrum Evil, Quackery and Allied Matters Affecting the Public Health*, vol. 2 (Chicago: AMA Press, 1921); Cramp, *Nostrums and Quackery and Pseudo-Medicine*, vol. 3 (Chicago: AMA Press, 1936).

5. Wilshire's private letters reveal that he was selling 400 belts a day in November 1926; that he sold 3,000 at the end of 1926, and that by April 1927 he estimated that he had sold 50,000. These numbers, which could be influenced by Wilshire's desire to appear more successful than he was, are largely corroborated by sales records kept by agents and secretaries in his offices. For the specific numbers see Wilshire to R. B. Howell, November 6, 1926, UCLA Wilshire Collection (LAWC), Box 5, Folder 7, Henry Gaylord Wilshire to the Council on Pharmacy and Chemistry, 12/16/25, 0403-11, American Medical Association's Historical Health Fraud Collection [HHFC], and Wilshire to Emile Berliner, April 5, 1927, LAWC, Box 5, Folder 9. For general sales figures see business records, box 7, folder 10, LAWC.

6. The Pasadena office, for example, reported selling twenty-six belts in March 1926 and administering 539 treatments. This figure suggests that, using a relatively conservative 1 to 4 ratio, roughly 200,000 people received in-store treatments. See business records, box 7, folder 11, LAWC.

7. They also chose the I-ON-A-CO's imitators, among them the Theronoid, created in 1928 by Philip Isley who had worked in Whilshire's Cleveland office. Other followers included the Electronet, Magnecoil, Iona-tone, and Restoro. For information on later versions of the I-ON-A-CO, see "Electromagnetism and Your Health" (Minneapolis: Theronoid of Minneapolis, ca. 1930); Electronet advertising letter, 7/13/25, Electrotherapeutics folders, 0229-08, AMAHHFC; letter on the Magnecoil from M. G. Ripley to N. B. Salerni, 3/23/28, ibid.; the report on the Restoro from the Chicago Better Business Bureau, 8/31/28, ibid., 0232-17. For information on medical practices in the early twentieth century, including the findings and impact of the Flexner report on American medicine in 1911, see G.J. Barker-Benfield, *The Horrors of the Half-Known Life* (New York: Routledge, 1999).

8. This connection to automobiles may have been part of the device's appeal. One customer wrote to Wilshire that she had heard his radio address, gone to the Lawrence, Kansas office, and had a treatment. She wanted to return the next day, but could not because she had to go to the auto show. To people like this "Mrs. Fey," the I-ON-A-CO's design may actually have appeared modern through this connection. See Fey to Wilshire, undated letter, box 8, folder 4, LAWC.

9. Wilshire began the design phase with a version that weighed seven pounds, pairing this down to two pounds by the time of production in 1925. See Wilshire to Mary McReynolds Wilshire, July 16, 1925, LAWC, Box 4, Folder 17.

10. This lack of visible electricity was remedied by the addition of a second coil that attached to the first. This smaller coil had no function other than to light up when the device was plugged in. One can see the coil in advertising materials. It is also mentioned in Wilshire's correspondence; see for example Wilshire to Macfall, March 31, 1926, box 7, folder 11, LAWC.

11. Haladane Macfall, a friend and agent of Wilshire's in England, found it difficult to sell the device because there was no way people could tell if it was working. See Macfall to Wilshire, March 14, 1926, box 7, folder 11, LAWC.

12. This article focuses on why people believed I-ON-A-COs cured rather than exploring whether or not they actually cured. There is evidence, however, to suggest that the device may have cured some ailments. Although electromagnetic currents are not sufficient to cure major diseases, the I-ON-A-CO's light electric massage, if applied directly to accupoints, could have remedied minor ailments like fatigue, hemorrhoids, goiter, earaches, stiff necks, and sore throats. Electric massagers continue to be marketed to ease tired feet and increase circulation. Electrodes, applied directly to affected muscles and tendons, are used by physicians to alleviate chronic pain and tendonitis. TENS or transcutaneous electrical nerve stimulation machines are available to laboring women in the United States and Europe to ease the contraction pain. With its low-level electromagnetic current, the I-ON-A-CO lacked the force of current medicinal devices. As a result, it did not have the ability to

relieve intense pain or permanently cure a condition; it may, however, have been sufficient in some cases to temporarily relieve a patient's suffering. For more information on the therapeutic value of low-level currents, see J.A. Chiu et al., "Transcutaneous Electrical Nerve Stimulation Helps Massage Parts," *Essential Inforamtion on Alternative Health Care* 5, No. 7 (November 1999):16.

13. There is not yet a critical biography of Wilshire or his wife Mary MacReynolds Wilshire. Brief accounts of Wilshire's career can be found in the celebratory account of his I-ON-A-CO written by George Duriand and in brief newspaper articles. The most complete record of Wilshire's life can be found in the unpublished biography written by his son, Logan Wilshire, in the 1960s, held in the UCLA Wilshire collection archives.

14. For information on Wilshire's early life, including his cousins, Joseph Wilshire, who was president of Fleishman's Yeast, and another cousin who was head of a coal company, see Wilshire's thin official biography, George J. Duraind, *Gaylord Wilshire and His Amazing Discovery* (Los Angeles: The Iona Company, 1927), available in LAWC. One can also find information on Wilshire's real estate speculations, including the founding of the city of Fullerton in Mike Davis, *City of Quartz: Excavating the Future in Los Angeles* (New York: Verso, 1990).

15. Wilshire's letters reveal that he encouraged his son, Logan, repeatedly during Logan's stay in Italy in 1926, to send an I-ON-A-CO to Mussolini with the hope that it would improve his health. He also claimed to have discontinued publication of *Wilshire's Magazine* in the 1920s because, as he told one supporter, he had only "destructive criticism" for the current state of American society. See Wilshire to Logan Wilshire, March 1926, Box 4, Folder 1 (check) LAWC and Wilshire to Edgar Lore, May 22, 1924 (box folder?) LAWC.

16. For more information on turn-of-the-century health reformers like Kellogg and Horace Fletcher see Harvey Green, *Fit for America: Health, Fitness, Sport and American Society* (Baltimore: Johns Hopkins University Press, 1986) Logan Wilshire's unpublished biography of his father contains further information on his father's health habits. See LAWC.

17. The Ex-cell-o bread company, for which Wilshire was a distributor briefly in 1925, probably provided the inspiration for the I-ON-A-CO name. For more information on Wilshire's radium projects and his interest in Ex-cell-o, see Wilshire to Prince Hopkins, June 10, 1924, LAWC; and Wilshire to Dr. Frank McCoy, January 27, 1925, box 4, folder 13, LAWC.

18. Wilshire to Dr. W. A. Newman Dorland, September 17, 1926, Box 5, Folder 5, LAWC.

19. See, for example, Wilshire to Mary McReynolds Wilshire, October 7, 1926, Box 5, Folder 6, LAWC.

20. Wilshire to Mr. Edgar H. Lore, May 22, 1924, box 4, folder 4, LAWC.

21. Wilshire mentions his arrest in his biographical data, LAWC.

22. See also "A Delightful Road to Health," *Liberty* (January 15, 1927): 82. Duraind, *Gaylord Wilshire*, 9.

23. Wilshire, "The Short Road to Health," box 8, folder 11, LAWC.

24. For more information on Mesmerism and its relation to pseudo-health cures see Thomas de la Peña, *The Body Electric*, especially the introduction.

25. "Short Road to Health," the original article by Warburg was a reprint of a talk given at the Rockefeller Institute for Medical Research printed as "Iron, the Oxygen-Carrier of Respiration-Ferment" in *Science* v. 61, n. 1558 (June 5, 1925): 575–582.

26. Wilshire spoke at numerous public events including the food show at the Ambassador in Los Angeles in 1925 and on radio shows broadcast in California, Washington, Oregon, Idaho, and Kansas. He also spoke regularly at his stores to potential consumers. See A. P. Fey to Wilshire, Box 8, Folder 4; R. G. Spaulding to Wilshire, November 20, 1926, Box 8, Folder 3; and Wilshire to J. A. Stevenson, April 28, 1925, Box 4, Folder 16, all LAWC.

27. Wilshire's friend Craig Sinclair, wife of Upton, urged him to drop the photo that she thought looked too much like "the fakers" with its cloak. Others thought he looked like George Bernard Shaw. See Craig Sinclair to Wilshire, March 5, 1926, box 4, folder 2 and A.B. Pottenger to Wilshire, October 26, 1926, box 5, folder 6, LAWC.

28. Wilshire to Macfall, May 7, 1926, Box 5, Folder 4, LAWC.

29. For letters chronicling Wilshire's financial situation in 1924 see Wilshire to Leffingwell, January 1, 1924, Box 4, January–February; Wilshire to Regenaldo, January 2, 1924, Box 4, January–February; Wilshire to John C. Benton, April 22, 1924, box 4, folder 3; Wilshire to Editor, *Irish Statesman*, July 8, 1924, box 4, folder 7; all LAWC. Wilshire appears to also have been considering other creative real estate options, such as the idea that he could turn his beach cottage into a beach club and charge memberships. For that scheme see Wilshire to Mary McReynolds Wilshire, August 21, 1924, box 4, folder 8, LAWC.
30. The Bakken Library of Electricity and Life has a collection of blood samples sent to Abrams for analysis. These attest to the device's popularity, as they are sent from numerous regions of the United States.
31. For an account see Eric Jameson, *The Natural History of Quackery* and Alan Gauld, *Electrotherapy in the United States* (Minneapolis: Medtronic, 1977), especially "the great patent medicine era."
32. For information on Wilshire's interest in Abrams and the Autoclast see Wilshire to Regenaldo, January 2, 1924, Box 4, January–February 1924 and Wilshire to Prince, January 21, 1924, Box 4, January–February, all LAWC.
33. Wilshire to Prince, January 21, 1924, B 4, January–February, 1924, LAWC.
34. See McIntyre to Wilshire, February 21, 1925; Wilshire to Sterling, February 25, 1925; and Sterling to Wilshire, February 28, 1925, all in box 7, folder 1, Wilshire collection, UCLA.
35. For information on Sinclair's endorsement, see Robert Ernst, *Weakness Is a Crime: The Life of Bernarr Macfadden* (Syracuse, NY: Syracuse University Press, 1991), 53. For further information on Sinclair, see Leon Harris, *Upton Sinclair, American Rebel* (New York: Crowell, 1975). For information on Sterling, see Holbrook, *The Golden Age of Quackery*, 141–142.
36. For Wilshire's suggestion that his friend Craig Sinclair use the I-ON-A-CO to avoid the surgery for breast cancer, see Wilshire to Craig Sinclair, March 5, 1926, Box 5, Folder 2.
37. This belief was shared by others who believed in the device. Wilshire's biographer, George Duriand, began their association by writing to Wilshire in 1926 asking to sell the device because of its "beneficient influence" that was creating a "great national affair." Duriand to Wilshire, October 26, 1926, Box 8, October 16–31, 1926, LAWC.
38. For a list of agents, which includes one physician, see "List of Iona Agents," Ionaco 7–12, 1925, box 7, folder 9, LAWC.
39. Some of the cities in which Wilshire advertised: Pittsburgh, Detroit, San Francisco, Cincinnati, Newark, Spokane, San Diego, Salt Lake City, Cleveland, and Omaha. See attachment to letter from Arthur Cramp to Dr. Fishbein, 10/22/28, 0913-10, American Medical Association's Historic Health Fraud Collection (AMA Headquarters, Chicago) AMAHHFC.
40. Distributors purchased the devices at roughly one-half the retail cost and then sold them to customers, requiring, at first, a large cash investment on their part.
41. Nave to Wilshire, October 18, 1926, box 5, folder 6; Sarles to Wilshire, October 17, 1926, box 8, folder 1, Hall to Wilshire, November 14, 1926, box 8, folder 3, all LAWC.
42. The classic text on this rapid change and the unsettling impact it had on Americans is Robert and Helen Lynd, *Middletown: A Study in American Culture* (New York: Harcourt, Brace, 1929).
43. See Fenton to Wilshire, January 27, 1926, box 7, folder 10, LAWC. We also know that Fenton may have once had more money, as she jokingly writes to Wilshire (whom she seems to know well) that "poor folk" like her can't buy the device.
44. Linsebarth to H. R. Learns, January 11, 1926, box 7, folder 10, LAWC.
45. The I-ON-A-CO nurse was probably an important symbol of the effectiveness of the product. Theronoid, an product identical to the I-ON-A-CO, was developed and promoted by Philip Isley, a one-time manager of Wilshire's Cleveland office. Its shipping box featured a prominent figure of a nurse sitting in a chair using the device with a smile in her face. Isley probably used the symbol because he recognized the importance of the female, sex-symbol/expert figure from his previous company experience. See the illustration accompanying the box at http://www.americanartifacts.com/smma/thero/advert/az411c.
46. See Mrs. B. Bredemus to Wilshire, October 6, 1926, box 7, folder 15, LAWC.

47. This same position of status may have been sought by people who were not demonstrators. One of Wilshire's employees described a gas station owner he had met in Portland, Oregon who was giving free I-ON-A-CO treatments with the purchase of a tank of gas or oil change in 1927. One imagines drivers pulling up to fill their tanks and stepping into his office for a quick treatment. Such a scenario likely made the owner define himself as part medical professional and part gas-station owner. See Carl Wernicke to Wilshire, January 22, 1927, box 10, folder 18, LAWC. There is also a letter from a scientist who claimed to know Otto Warburg asking to be employed selling the device because of his skills as a speaker and his credentials as a researcher. See August Alger to Wilshire, October 31, 1926, box 8, October 16–31, 1926, LAWC.. Research also needs to be done on the physicians who agreed to prescribe, and in some cases, distribute the I-ON-A-CO. Drs. Frank McCoy and J. Byron Sloan of Los Angeles became product boosters, as did David Starr Jordan of Stanford. Annie Lyle of San Francisco also seems to have supported the product, initially, though she denied this later. Further, physicians like Rodney Madison worked for I-ON-A-CO on staff, prescribing treatments and evaluating cures. For information see David Staff Jordan to Wilshire, April 9, 1926, box 7, folder 11; McCoy to Wilshire, March 10, 1926, box 5, folder 2; Madison to Wilshire, October 29, 1926, box 8, Ionaco October 16–31; Sloan to Dr. H. F. Fisher, March 27, 1926, box 5, folder 2; Lyle to Wilshire, May 3, 1926, box 8, folder 6, all LAWC.
48. Wilshire's records indicate centers in Oakland, Los Angeles, San Francisco, San Diego, Berkeley, Long Beach, Chicago, Portland, Seattle, Tacoma, Denver, Kansas City, Palo Alto, Alhambra, and Cincinnati. See for example, Spencer to Wilshire, October 17, 1926, box 8, folder 1 and J. R. Owens to K. E. Vankuran, October 21, 1926, box 8, folder 1, LAWC.
49. The estimate can be found in Wilshire to Signor Odon Por, October 7, 1926, box 5, folder 6, LAWC.
50. In Oakland the office was in the Ray building and in Seattle the Shafer building. Wilshire was considered by at least some architects as an important supporter of their work. This may be due, in part, to his placement of I-ON-A-CO offices in architecturally significant sites. Rudolph Shindler wrote to Wilshire asking him to consider taking him on as an architect for his business ventures. See R. M. Shindler to Wilshire, March 18, 1924, box 4, folder 2, LAWC.
51. In July 1926 Wilshire wrote to his son Logan that the San Francisco office was selling 1,250 belts a month and had become his main center of business. Wilshire to Logan Wilshire, July 20, 1926, box 5, folder 5, LAWC.
52. In a two-page advertisement in the Seattle *Post-Intelligencer* on October 31, 1926, this illustration was expanded into an actual photograph featuring the high-rises of what it described as the "famous Wilshire district of Los Angeles." The caption, under the illustration, further emphases Wilshire's connection to modernity by describing him as "the first to start" a movement for a "new architectural age in the United States" through his "world-renowned Wilshire Boulevard." See section 1, pp. 20–21.
53. For information on radio shows see R. G. Spaulding to Wilshire, November 20, 1926, box 8, folder 3 and Mrs. A. P. Fey to Wilshire, 1926 (undated), box 8, folder 4, LAWC. Even after Wilshire's death, radio stations continued to broadcast his shows. Allen Shoenfield talks about radio broadcasts in March 1928 in a letter to Morris Fishbein, Wilshire correspondence, AMA archives, 0914-04. For information on billboards see A. B. Pottenger to Wilshire, October 26, 1926, box 5, folder 6 and Ida Hendrichson to Wilshire, October 12, 1926, box 5, folder 6, LAWC. Motion picture advertising is discussed in Neuenberg to Wilshire, October 26, 1926, box 8, October 16-31, 1926. The food show is discussed in Wilshire to J. A. Stevenson, April 28, 1925, box 4, folder 16, LAWC.
54. This may have been emphasized, as well, for those who received Wilshire's promotional flyer, appropriately named "The Ionaco News." In 1926, 200,000 were printed for mass-mail distribution in cities with I-ON-A-CO offices. See H. A. Andrews to Wilshire, October 19, 1926, box 8, Ionaco October 16-31, LAWC.
55. Wilshire to Mollie Price Cook of New York, April 25, 1925, box 7, folder 1, LAWC.
56. Wilshire promotional brochure, box 10, folder 21, HGWC.

57. For an account from one of these "nurses," who details the tasks in her job, including setting out the I-ON-A-COs, greeting customers as they arrived, and overseeing their treatments, see Mrs. Bredemus to Wilshire, October 6, 1926, box 7, folder 15, LAWC. Patients may also have been made more aware of the power of that electricity by demonstrations done in treatment rooms. Wilshire often advised his agents to show prospective customers "the magnetic test," which involved taking a steel hair pin and laying it on the inside of the belt, thereby magnetizing it (presumably as the blood would be magnetized) and enabling one to then "pick it up with a screw driver or a nail file." See Wilshire to Macfall, March 17, 1926, box 7, folder 10, LAWC.

58. It is difficult to know whether this context did, in fact, sell belts. We do know, however, that Wilshire believed that his free demonstrations (what this photograph probably illustrates) did drive sales. "We give a number of free treatments and during these periods make many sales," he wrote to his sales associate W. H. Leffingwell, October 26, 1926, box 8, folder 1, LAWC. These sites may have even influenced people who purchased I-ON-A-COs for home use: the same letter states that people often came and visited the offices when they made payments on I-ON-A-COs purchased by mail.

59. For a copy of this form, see The Iona Company Clinical-Sales Record, box 8, folder 12, LAWC.

60. It may have also made them feel more in control of their bodies than did regular medical treatments. Wilshire realized in 1927 that people were often more likely to purchase I-ON-A-COs once they had met with representatives of his medical department (typically physicians who were employed by Wilshire to promote the device). "Ionaco owners should be encouraged to come to our offices for examination and consultation," he reported, citing the "psychological value of this co-ordination of medical knowledge to the Ionaco" as a key factor in increasing sales. Internal memo from Wilshire, February 14, 1927, box 8, folder 6, LAWC.

61. The I-ON-A-CO's use of women in its advertisements distinguishes it from earlier electric belt products that advertised primarily using male figures. I have found only one advertisement that featured a man, from the double-page ad in the Seattle *Post-Intelligencer*, October 31, 1926. Here an illustration shows a male figure throwing a life preserver to rescue a man out in the water. The drowning subject reaches out for the object, labeled the I-ON-A-CO Magnetic Life Buoy. See section 1, pp. 20–21.

62. It is perhaps significant that the Fresno letterhead, produced after the "Short Road" brochure, uses a much larger light than does the original brochure. Promoters may have realized over time that many consumers were interested in seeing the connection between the body, the device, and the warm lamp of home.

63. For example, when agents began the Portland office, they received a list of residences that were customers of the Electric Power and Light Company and were able to conclude that of the 5,700 people in the town, 2,500 to 3,000 received electric service and were therefore potential customers. See Carl Wernicke to Wilshire, February 2, 1927, box 8, folder 6, LAWC.

64. For accounts on electric connections and testing techniques see "Short Road to Health," 20 and "Wilshire's I-ON-A-CO, box 7, folder 9, Ionaco 7–13, 1925, LAWC.

65. Wilshire's device was designed to light up only with alternating current because as Wilshire reported, 96 percent of U.S. electricity was 110 volt AC. With DC currents, which were used in New York, the belt would not light up. Wilshire's letter is found in box 7, folder 10, Wilshire to Wilshire, January 25, 1926; Abbott's is Abbott to Wilshire, January 6, 1926 and January 16, 1926, box 7, folder 11. For information on Wilshire's guess as to the percentage of AC users, see Wilshire to Macfall, March 31, 1927, box 7, folder 11.

66. For more information on the cultural change that accompanied electrification, see David Nye, *Electrifying America: Social Meanings of a New Technology* (Cambridge: MIT Press, 1990)

67. Barton used the Theronoid, a leading I-ON-A-CO imitator. Annie E. Barton to Mrs. J. C. Bitterman, April 14, 1932, at http://homepages.rootsweb.com/~barton/annie/year1932.

68. For an analysis of the evolution of the home from a productive, functional site to one of symbolic, protective value, see Clifford Edward Clark, *The American Family Home: 1800-1960* (Chapel Hill: University of North Carolina Press, 1986).

69. Internal memo, February 14, 1927, box 8, folder 6, LAWC.

70. Wilshire to Jordan, March 26, 1926, box 5, folder 2, LAWC.
71. Clippings found in box 15, folder 16, LAWC. For information on Mary's courses, including syllabi and outlines, see box 15, folder 11, LAWC. The holdings at UCLA, part of the Henry Gaylord Wilshire's collection, include Mary's manuscript, her letters to friends and family, and her own dream analysis journal kept over a period of several years.
72. Clippings found in box 15, folder 16, LAWC.
73. Manuscript, box 15, folder 12, LAWC.
74. This figure comes from Mary's manuscript in which she claims to have "scientifically interpreted" more than fifty thousand dreams. We know that Mary was still practicing analysis in 1924 when Wilshire began developing the I-ON-A-CO because of a letter from Alfred Bailey to Mary detailing their interaction that year in Honolulu. In the letter, Bailey thanks her for allowing him to read her manuscript and asks advice about further readings in practical applied psychology. This seems to be the same Dr. Alfred Bailey who promoted the dangerous pseudomedicine "Radithor" in the 1920s, though I have been unable to confirm that connection. See Bailey to Mary Wilshire, October 31, 1924, box 4, folder 10, LAWC.
75. It is possible that this connection may have been behind efforts Wilshire made to style his own portrait on promotional materials to look like Freud.

3

Technology and Disability

JIM TOBIAS

Abstract

Technology products can provide support for some functions that people with disabilities want to perform. These are either specialized assistive technology (AT) products, designed just for people with specific functional limitations, or are mass-market products with certain accessibility features, sometimes referred to as universal design (UD). Especially in the realm of information and communication, these two categories cooperate and contend with each other technically, politically, culturally, and economically.

Assistive Technology

Rudimentary devices intended to assist people with disabilities have existed in most cultures and eras.[1] The rigorous and focused application of technology for this purpose is a characteristic of modern industrialized cultures. Assistive technology (AT)[2] has become part of many U.S. government programs at all levels, with large-scale programs in place to research and develop new technologies,[3] and to provide direct services and products. For example, schools are required to provide assistive technology to students with disabilities, employers receive tax benefits for workplace accommodations, and federal agencies must purchase electronic and information technologies that are accessible.

Businesses have been encouraged to enter this market[4]; rehabilitation professionals have been encouraged to add technological abilities to their practices[5]; people with disabilities have been encouraged to consider how they could benefit from using AT.[6] The most highly targeted activity areas are education and employment, but communication, transportation, and recreation have also received significant attention.

The main developmental phase of AT in the United States occurred in the 1970s and 1980s. This period saw the establishment of federal research laboratories, an increase in public funding for rehabilitation services of all kinds, and the origin of small AT firms specializing in electronic-based products. The main professional organization, RESNA (Rehabilitative Engineering and Assistive Technology Society of North America), was founded in 1979.[7]

This era may have been a unique confluence of technological and cultural forces. Electronic technologies were maturing, computer capabilities first reached a mass market, and medical techniques managed to treat and stabilize a large proportion of previously fatal spinal cord and head injuries. At the same time, people with disabilities began their civil rights movement, at times spearheaded by Vietnam War veterans with disabilities. Beset by a popular crisis of faith in technology brought on by environmental awareness and antimilitary skepticism, American technologists could point to assistive technology as human-centered and benevolent.

Although there is a powerful and diverse AT armamentarium,[8] few people with disabilities (estimates for some important product categories range around 10 percent) actually receive its benefits. Scarcity of information about assistive technology, a weak local services infrastructure, and high costs are usually identified as key barriers .[9] Also referred to is the stigma attached to AT, since the clinical appearance of much of it immediately marks the user.[10] One elder with low vision refers to the AT products paraded before him by his concerned children as "chrome plated rat-traps"; he refuses to use any of them although it means that he must forego certain activities in which he used to participate. In his case, he no longer reads a daily newspaper, and clearly expresses frustration about that.

Universal Design

Beginning in the 1960s it was recognized that buildings can impose a pattern of barriers to people with disabilities, especially people who use wheelchairs. Although the first set of responsive architectural standards was published in 1961, it took the 1968 Architectural Barriers Act to require accessibility in federally funded construction. As these and other laws and regulations took hold, "it became apparent that segregated accessible features were 'special,' more expensive, and usually ugly. It also became apparent that many of the environmental changes needed to accommodate people with disabilities actually benefited everyone. Recognition that many such features could be commonly provided and thus less expensive, unlabeled, attractive, and even marketable, laid the foundation."[11]

Accessibility advocates and technologists began to recognize the extensibility of universal design (UD) beyond the field of architecture. In part as a response to some of the weaknesses of the AT delivery system, disability advocates, designers, and policy specialists joined the movement for *universal design*[12] (UD), also called "design for all." The goal of universal design is to include as much accessibility as possible in mainstream products and services, rather than require users to acquire and use assistive technologies. Thus an appliance with a large character display or a speech synthesis capability would meet the needs of many people with impaired vision, obviating the need for a special AT device, a portable electronic magnifier. Examples of the contract between AT and UD products are shown in Figure 3.1A-D. Note that the word *universal* is somewhat misleading, as no one claims to be able to design a single product that will serve the needs of every imaginable user. UD is often referred to as a process of continuously expanding the potential range of users.

UD also expands the working concept of disability beyond the clinical definition to include people without disabilities who are in disabling situations. People using strollers or hand trucks benefit from curbcuts just as wheelchair users do; closed captioning is probably used as much in loud sports bars as in the homes of people with hearing loss. This further reduces any stigma attached to using an accommodation.

UD is a response to AT's medical orientation. Instead of fixing the person with a disability by equipping him or her with an accommodation that fits the way the world is currently designed, UD identifies the locus of disability as the unnecessarily inaccessible built environment. The origin of disability is not within the "broken" person, but at the point of intersection between the person and the built environment. That is, disability becomes the gap between what a person can perform and what the environment demands. The problem is "out there." The elder with low vision referred to above uses his daughter's AOL (America Online) account to listen to news stories over the telephone through a service called AOLbyPhone. Although it does not replace reading a complete newspaper, it does satisfy his curiosity, is easy to use, and gives him more control than listening to news radio.

In the light of UD, disability becomes not a tragedy but an ergonomic situation. All humans exist as points within a set of bell curves that indicate their differing abilities. All consumers are confronted by technologies that are usually or occasionally difficult or impossible to use. UD claims to be able to expand the potential market of any product through increased usability. In fact, UD advocates often make sweeping claims about its revenue potential.

AT and UD in Sociocultural Context

It is not enough to consider AT and UD as two technological approaches to the needs of people with disabilities. As Rudi Volti puts it, "[i]n

 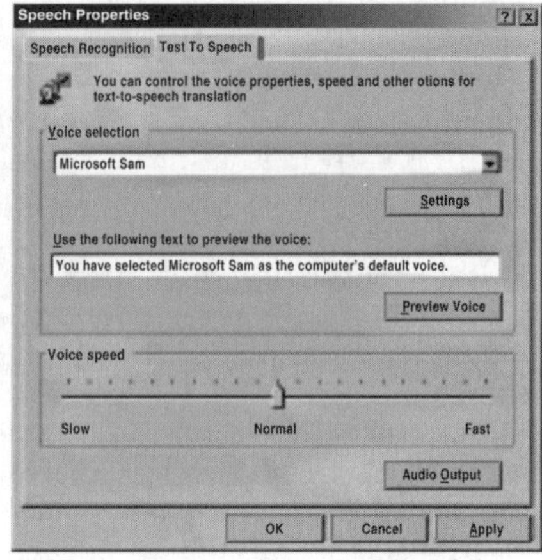

 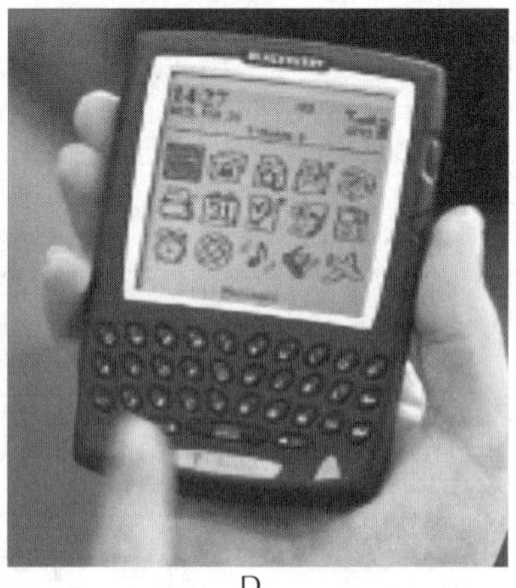

Fig. 3.1 (A) A standalone speech synthesizer sold in the 1980s for about $500, had one voice, and required special programming. (B) Today's operating systems offer free software-only speech synthesis, with many built-in voices. (C) A TTY requires identical equipment at both ends of the call and cannot connect with computers or perform computer functions. (D) The Blackberry is a mainstream wireless text messaging device with other functions; it often sells for less than a TTY. (Images courtesy of Inclusive Technologies.)

considering the influence that technology in general or any single technology has over human affairs, it is therefore necessary to consider not only the technology and its presumed 'imperatives,' but also the key human agents of the technology, the organizations in which they operate, and how these influence the course of technological change."[13] AT and UD represent distinct expressions of certain cultural values about technology and certain professional and commercial motivations.

Technological Utopianism [14] [15]

AT and UD both bask in the glow of a generalized uncritical belief in the benefits of technology. This has raised unreasonable expectations and prevented reasonable assessments of technology outcomes compared with other interventions. Technological interventions are automatically deemed superior, and can be improved only through the use of more improved technologies.[16]

Connected with technological utopianism, especially in its American form, is a focus on technology as a provider of absolute independence. In the American dream, everyone has a fuel-cell-powered, all-terrain vehicle with a wireless Internet connection and a robotic fast-food package grabber. In the rehabilitation version, the vehicle is an all-terrain wheelchair. Rather than ensuring equality of social participation, accessible technology often appears to reinforce the cultural value of omnipotent isolation. An education professional who is blind used to use interns and graduate students to read his mail to him. For several years he used a scanner and screen reader program to read his mail independently. Even though accuracy was a problem and the system was none too easy to operate, he felt it was important to advance his technological sophistication and address his secondary need for privacy. However, he recently returned to using readers instead, both because it is easier and more friendly.

AT and UD have also been put forward as an example of humanistic technology, technology in service to humanity, especially when technology is under attack (e.g., antiwar, environmentalism). Student projects, volunteer projects and organizations, military and NASA spin-offs characteristically emphasize the "promise of a new life." Breathless press accounts of breakthroughs in accessible technology reinforce this cultural value of redemption through computation. AT has for thirty years shadowed all the hot trends in mainstream technology: speech synthesis and recognition, CB radio, robotics, virtual reality, and the Internet.

Information technology corporations point to the accessibility of their products as evidence of the humaneness of both their products and their cultures. In fact, in the late 1990s when technological job openings outnumbered applicants, some companies showcased their accessibility work in recruitment sessions for new engineering graduates. An idealized image of the technology — omnifunctional, hyper-personalized, rapidly advancing, increasingly inexpensive, and instantly usable — reinforce the image of a market that offers an irresistible bounteous flow, sweeping away inconvenience and inability alike.

Technological Idealism and Totemism

In the context of technological utopianism, tokens of accommodation are coin of the realm. Rather than measure the extent to which AT or UD actually improves the lives of large numbers of people with disabilities, policymakers and others have been willing to see a promising AT prototype or a single universally designed mass-market model as an indication that the problem is solved. This focus on technical innovation rather than massive implementation has discouraged efforts directed at solving information and service infrastructure problems. Although lack of information, professional training, and local support services have been repeatedly identified as key barriers to the rollout of accessible technology, these issues play poorly; they lack glamour. The Potemkin village aspect of the exhibit halls of the major accessibility conferences allows too many visitors to walk away — or roll away — reassured, relieved, and refreshed. One AT vendor, himself disabled, likes attending trade shows because their scope makes him feel as if disability is finally being paid sufficient attention. He "count[s] the number of exhibits as an index of how seriously society takes us."

In another more speculative sense, AT serves to mediate how disability is viewed. Without AT, a person with a disability may be seen as "strange," "wild," or completely external to the range of human abilities. With AT, the same person may still be perceived as different — indeed, as was said above, AT can be the very chrome-plated mark of disability — but no longer external to humanity. After all, what other entity not only makes and uses tools, but also carries them as a badge of capability, competence, and wealth?

Contending Models of the User and the Practice

How do AT and UD envision their users and their practices? Their contention in this arena carries many of the same cultural themes found in other domains.

AT appears to offer, and indeed often delivers, a fully personalized service that addresses the technical needs of a person with a disability, at little or no cost. In its best incarnation, the AT model subsidizes the services of a range of well-trained professionals, who evaluate the client, select the most appropriate product, and deliver it with the promise of perpetual maintenance and upgrading. It is no exaggeration to say that hundreds of thousands of Americans have received billions of dollars in professional services and products that they could not have found, selected among, or paid for by themselves.

On the flip side, AT envisions its users firmly within a disability services network. In order to learn about, be evaluated for, or receive funding assistance for an AT product, the potential user normally has to enter that network, such as by becoming a client of a public assistance treatment program. He or she may also have to tolerate labeling, the lesser autonomy of a client/patient, and other sociocultural phenomena associated with such service networks.

Rehabilitation professionals are predominantly nontechnical by proclivity; they are now expected to perform some highly technical AT functions. *Rehabilitation engineering*, the engineering profession's term for AT and UD, has had to establish itself as a separate practice, apart from therapeutic professions, other branches of engineering, and general usability. This led to the development in 1979 of a professional organization (RESNA) and a push for professional standards and licensure. AT companies are generally too small to employ many rehabilitation engineers; most of them are employed in research institutions, such as the federally funded Rehabilitation Engineering Research Centers (RERCs).

UD sees its users within the mass market, not in a disability niche. Ideologically at least, UD posits a more empowered person with a disability. Like any other consumer, the person with a disability participates in a mass market, seeking the best match between his/her abilities and needs, and the range of products and services that can serve those needs without demanding too high a level of human performance.

On its flip side, however, UD can be quite unpredictable. Although mainstream products are legion and ubiquitous, it is rarely clear whether a given class of products or a given model actually will meet a consumer's specific need. A feature like the audible low battery indicator, important for cell phone users who are blind, is too low on the feature list to make it onto the box or into the consciousness of a salesperson.[17] An accessible product, information about which is excitedly passed from person to person, suddenly is replaced by an inaccessible upgrade.[18] Often the model with the widest range of features is also the most expensive model; people with disabilities invariably pay more, often for features they cannot use.[19] Within these companies there are many professions contending for control of the accessibility program, which touches departments of public relations, legal/regulatory/governmental affairs, human factors, quality assurance, marketing, and customer support as much as it affects design and engineering.

Accessibility arises at the same time that mainstream industry has tended to displace engineers from early product development decisions; companies claim to be market-driven rather than technology-driven. In response, engineers in some companies have attempted to seize upon universal design as an open space for the pure practice of their profession, safe from the incursions of marketers.

Compatibility

Compatibility is supposed to provide a neutral meeting ground between AT and UD. Technically, compatibility means that enough standards of interconnection and intercommunication are adhered to so that the AT device and the mainstream product appear to the user as unified and complete in operation. Behind the technologies, however, hide a host of hobgoblins. If standards are to be used, then whose? Where is the dividing line of responsibility?

A common problem with computer accessibility is that changes to the operating system may require matching changes in the AT. It may take a resource-poor AT company six months or more to catch up with the change; operating system companies are reluctant to include AT companies in the select group of partners entitled to prerelease technical consultation .[20]

There is some audible interference between digital wireless phones and hearing aids; this is annoying enough to prevent many hearing aid wearers from using modern cell phones. Are the phones broadcasting too powerfully, or are the hearing aids not immune enough? Resolving this issue took over five years of disputation and innovation, eating up tens of thousands of hours of the expert staff time of advocates, researchers, regulators, and engineers alike.[21] [22]

Economic Competition between AT and UD

The competition between AT and UD is similar to that between a boutique and a big box discount chain. People with disabilities are the principal clientele of AT, resulting in excellent service but at scarce locations and high prices. Mass-market companies believe that any additional revenue from this market segment may not be worth the cost of modifying their products or business practices. They tend to make the easiest changes and offer them on a take-it-or-leave-it basis.

Mainstream companies are not as well motivated to serve their customers with disabilities as AT companies are. For example, one study showed that service calls to five out of six television manufacturers failed to result in accurate information about adjusting captions, while all closed caption decoder manufacturers (three AT companies) answered the same question accurately.[23]

One ongoing concern is whether advancing mainstream technology jeopardizes the AT industry. Screen magnification programs offer computer users with low vision sophisticated ways of enhancing the visual display. However, most operating systems now contain rudimentary screen magnification utilities. Although their functionality is not as high as that of the specialty programs, the difference in cost (effectively zero for the operating system utility) has no doubt reduced the sales of screen reader software. When AT manufacturers first made screen readers, they bundled together special word processing software along with the computer hardware and speech synthesizers for their customers who were blind. But soon these users demanded access to regular word processing programs so that they could share documents with others. This eliminated the value of the special software. Then multimedia computers with built-in speech synthesis capabilities came along, further eroding the added value in the specialty market. Today's screen readers are little more than navigational interfaces and software detours that safely reroute text from the screen to the audio card. What does the future hold for screen reader companies? One policymaker with a disability expresses scorn for the concept of universal design and has managed to include regulatory provisions that favor AT companies.

Similar effects are seen among deaf communicators. Previously relegated to specialty TTYs (essentially keyboards and small displays with no computing capability), more and more deaf people are using e-mail, instant messaging, and wireless text devices. There are some data that indicate an actual decline in TTY traffic as people shift to these other modes, which do not require special technologies on either end. In response, the main TTY manufacturers have begun to focus on other ways of serving their customers. Several technologically sophisticated TTY users state that they never initiate TTY calls any more, but must have a TTY handy in case they receive a call from someone less advanced.

Were it not for consumers' demand for access to mass-market products, mass-market companies might have gone on forever ignoring people with disabilities. But that consumer demand was translated into laws and regulations intended to force mass-market companies to pay attention.[24]

Public Policy

Even before the Americans with Disabilities Act (1990), a string of accessibility laws and resulting regulations aimed at improving the opportunities for full social participation by people with disabilities. Although few of these laws targeted technology directly, in many cases technology became a logical and effective implementation tool. That is, a school mandated to provide a student with a disability with a "free and appropriate public education"[25] often found that a measure of assistive technology was one effective way to provide it. The same was true of employers seeking to hire people with disabilities. These laws required accommodations, but did not specify which accommodation to use. There was no specific mandate for mass market products, and AT companies were happy to fill the need wherever they could.

However, beginning in 1996 with the revisions to the Communication Act, a new trend began that aimed expressly at mainstream technologies. Section 255[26] of that act requires telecommunications manufacturers and service providers to guarantee that their products and services are "accessible to and usable by" people with disabilities. In 1998 an amendment to the Rehabilitation Act[27] required federal agencies to purchase only accessible electronic and information technologies. This one-two punch of supply-side and demand-side legislation was a victory for UD activists, making widespread accessibility the law of the land for information technologies. The new regulations fostered new technical and market opportunities into which, it was hoped, large and small companies would be both pushed and pulled. But it also posed a problem.

On the one hand, UD and disability consumer advocates expected full accessibility, right out of the box, of all mainstream products. At the same time, however, they realized that compatibility with AT might be a more achievable goal, at least during an interim period. They were suspicious of the federal government's dedication to enforcement, doubted the ability and will of mainstream companies to maintain full accessibility over time, and appreciated the higher attention-share that their issues received from AT companies. As of this writing, the main enforcement trend appears to be moving toward requiring mass market products to be compatible with AT, except where some obvious full accessibility opportunities exist, where advocacy pressure has been strongest, or where AT is unavailable or ineffective.

This arrangement suits both AT and mass market companies. As long as they can guarantee compatibility, mainstream companies can safely avoid distracting, costly changes to their products. And as long as mainstream products stay on their side of accessibility, AT companies are guaranteed a stable market and a more or less monopolized channel of communication to consumers with disabilities.

It is, in fact, the UD champions, including those within mainstream companies, who threaten to upset the applecart. Equally motivated by technological utopianism's dreams of humanistic technology and the effective case disability advocates make for continued progress, they insist that the ever-advancing capabilities of new technology products be directly applied to meet the needs of customers with disabilities. They are setting the stage for new laws and regulations with expanded scope and clarified regulations.

Conclusion

Technology and disability display a remarkable interaction, fostering a lively emerging field that plays on and reinforces certain social, economic, and cultural phenomena while creating an array of interesting, useful products and features.

As social and technological practices, assistive technology and universal design demonstrate continuing trends as well as rapid evolution. Although they share certain characteristics, it is unlikely that these two will ever be reconciled for long: they are supported by essentially conflicting economic tendencies and clearly divergent images of their target populations.

In the world of AT, professional evolution has awarded decision-making power to therapists and rehabilitation counselors, although the across-the-board decline in public support for services has reduced their scope of action. In UD, regulations that were intended to force mainstream technology companies to radically redesign their products have been slow in accomplishing this goal, but have consolidated some of the key accessibility technologies that were easy to implement.

It is unfortunate that the large majority of people with disabilities remain at such a distance from this contention. In fact, they are still at too great a distance from the fruits of AT and UD themselves. There is little cause for optimism for the arrival of an energized disability community seizing the technology decision-making process. In recent years even some disability advocacy organizations have retired from direct technological battle, appearing both exhausted by their efforts to create current regulations and incapable of pressing the matter further without expensive technological resources of their own. Perhaps the most we can hope for is an increasingly informed consumer segment, with or without the goad of strong disability self-awareness.

Notes

1. Nunn, J. *Ancient Egyptian Medicine* (1996).
2. "Assistive Technology (AT) is any item, piece of equipment, or product system, whether acquired commercially off the shelf, modified, or customized, that is used to increase, maintain, or improve functional capabilities of an individual with a disability." Assistive Technology Act of 1998; IDEA Amendments of 1997, 20USC 1401(1), Sec. 300.5.
3. The National Institute on Disability and Rehabilitation Research (NIDDR) (http://www.ed.gov/offices/OSERS/NIDRR/) supports a network of seventeen Rehabilitation Engineering Research Centers (RERCs), each performing research and development on a separate, prioritized topic. NIDRR also supports dozens of other centers and field-initiated research on technology and disability.
4. NIDRR also supports commercialization of accessible technologies, both through its Small Business Innovation Research program and an active technology transfer program.
5. There is a specialized association for assistive technology professionals, RESNA (www.resna.org), as well as technology skill-building programs in professional associations of occupational therapists, speech-language therapists, special educators, and vocational counselors, among others.
6. In addition to government-funded consumer outreach projects, the major disability advocacy organizations promote accessible technologies to their members and stakeholders.
7. RESNA, *RESNA History*, 2003. http://www.resna.org/resna/history.htm
8. In addition to hundreds of models of general-purpose manual and powered wheelchairs, there are seventeen "beach wheelchairs" and eighteen wheelchair drink holders. ABLE-DATA: www.abledata.com, 2002.
9. *Federal Policy Barriers to Assistive Technology*, National Council on Disability, 2000, Washington D.C.
10. Pippin, K., Fernie, G. R., "Designing Devices That Are Acceptable to Frail Elderly: A New Understanding Based Upon How Older People Perceive a Walker. *Technology and Disability* 7, nos. 1–2, (1997): 93–102. Also Gitlin, Laura N., "Why Older People Accept or Reject Assistive Technology." *Generations, Journal of the American Society on Aging* 19, no. 1.

11. What is Universal Design: History and Background. Center for Universal Design, North Carolina State University. http://www.design.ncsu.edu/cud/univ_design/udhistory.htm

12. The term *universal design* was coined by the late Ron Mace, an architect with a disability, who was Director of the Center for Universal Design at North Carolina State University. One authoritative definition is "the process of creating products (devices, environments, systems, and processes) which are usable by people with the widest possible range of abilities, operating within the widest possible range of situations (environments, conditions, and circumstances)," "Universal Design Research Project, final Report," Trace Center, University of Wisconsin.

13. Volti, R. *Society and Technological Change*, 3rd ed., New York: St. Martin's Press, 1995, p. 257.

14. Segal, H. P. *Technological Utopianism in American Culture*, Chicago: University of Chicago Press, 1985.

15. Kling, R. "The Dreams of Technological Utopianism," in *Computerization and Controversy*, C. Dunlop and R. Kling, eds. Boston: Academic Press, 1991, pp 14–81.

16. Two examples come to mind. The telecommunications relay service (TRS, or *deaf relay*) is the subject of many experiments with speech technology, wireless communications, and IP telephony, but its most frequent users care more about relay operator language and typing skills. A steady stream of funding has been poured into robotics research for people with mobility impairments, but most of these people would prefer an improved system for personal attendant services.

17. Market Monitoring Report, U.S. Access Board, 1999. (http://www.access-board.gov/telecomm/marketrep/db_top.htm).

18. A talking pager was briefly on the market in 1998 and 1999. When it was withdrawn, its users with visual impairments had no easy time replacing the functionality they had become dependent on.

19. A recent issue is the use of wireless text messaging. Some deaf users have to purchase a full-price wireless voice account, then pay extra for the text messaging service, but receive no rebate even though they never make voice calls.

20. For two typical statements of this concern, see http://www.abilitech.org/AT/97spring.htm, September 17, 2003, and the remarks of Cheryl Cumings in the National Council on Disabilities' *The Accessible Future Release*, held June 21, 2001, National Press Club.

21. *In the Matter of Section 68.4(a) of the Commission's Rules Governing Hearing Aid- Compatible Telephones*; Federal Communications Commission, WT Docket 01-309 RM- 8658.

22. Note that this issue is less important in Europe, where the potential for interference is just as high. While American consumers pay for their own hearing aids, European health care systems pay publicly. Thus it was feasible to replace the interference-prone hearing aids of any person wishing to use a digital wireless phone.

23. Tobias, J., *Closed Captioning Customer Service Study*, Inclusive Technologies, 1998.

24. Thus we have the somewhat amusing insistence by advocates that even though UD dramatically enhances mainstream revenue, mainstream companies must be forced by law to adopt it.

25. U. S. Congress, *Individuals with Disabilities Education Act*, Part B (34 CFR Parts 300 and 301 and Appendix C).

26. The Federal Communications Commission (FCC) has jurisdiction for enforcement of this section. The FCC has a Disability Rights Office, found at http://www.fcc.gov/cgb/dro/.

27. Section 508 of the 1998 Amendments to the Rehabilitation Act, which is explained well on the federal website www.section508.gov.

II
Fixing Food

4

The Nutritional Enrichment of Flour and Bread: Technological Fix or Half–Baked Solution

MICHAEL ACKERMAN

When President Franklin Roosevelt famously contended during his Second Inaugural Address in January 1937 that "one-third of a nation [was] ill-housed, ill-clad, [and] ill-nourished," the high percentage of misfortune that he cited was neither hyperbole nor mere conjecture. The figure of one-third likely came from the results of a dietary survey that had recently been conducted by the U.S. Department of Agriculture (USDA). By comparing food consumption data to newly formulated quantitative nutritional requirements, USDA researchers had found that 35 percent of American families were consuming less than the minimum recommended amount of at least one essential vitamin or mineral, while just 27 percent had diets that provided a liberal amount of every nutrient that had been studied.[1] As the country drifted closer to war during the next few years, government leaders became increasingly troubled by this discovery, since health officials contended that nutritional deficiencies undermined the nation's defense capability.[2]

In response, health authorities, most of whom were physicians, sponsored an initiative that they predicted would overcome a significant part of the malnutrition problem. They recommended that all white wheat flour be enriched with the mineral iron and three synthetic B-complex vitamins: thiamine, riboflavin, and niacin. Each of these four substances was deficient in the American diet; although they all occurred naturally in the bran and germ of wheat, each of them was largely removed when the grain was milled to produce white flour. Incorporating these nutrients into white flour at the mill was thought to be an effective and relatively inexpensive way of increasing their intakes. Through the clever use of the federal food law and the enactment of mandatory enrichment legislation at the state level, nearly all white flour purchased in the United States was, by the end of the war, of the enriched variety and this has remained the case ever since. Similar actions were later taken to enrich corn meal and white rice, which suffered similar nutritional losses during processing.[3]

Yet not everyone endorsed this technological approach to fighting malnutrition. In fact, some of the country's leading nutritionists were the most vocal critics of enrichment. They argued that enriched white flour was nutritionally inferior to darker flours containing more of the natural vitamin and mineral content of wheat, and they recommended that enrichment be scrapped in favor of an educational campaign designed to persuade Americans to eat high-extraction wheat breads in place of white bread. This article will show how differences in professional values between nutritionists and physicians shaped the controversy surrounding the cereal enrichment program in the late 1930s and early 1940s. It will conclude with an assessment of the program's effectiveness.

Differences of opinion on the healthfulness of bread were not new in Depression-era America. One hundred years earlier Sylvester Graham, a clergyman turned health lecturer, achieved both fame and notoriety as a critic of white bread. Graham, who began his new career in the midst of a cholera epidemic, had come to believe that all disease was caused by excessive stimulation of the gastrointestinal tract, and he told his listeners that they could avoid disease by consuming a moderate diet of whole-wheat bread, vegetables, fruits, and nuts, and by giving up all foods — meat, spices, tea, coffee, alcohol, and white flour — that caused too much irritation to the alimentary canal. Graham especially condemned the use of bolted flour from which the outer bran of the grain had been removed, because he feared that the resulting highly concentrated flour was extremely stimulating. He also maintained that God had placed the proper combination and amounts of all the substances needed for human nutrition in the whole grain, and that man did not know enough about diet to compensate for the loss of any of these nutrients by the substitution of other foods. In addition, Graham accused commercial bakers of jeopardizing the public health because they added chemical adulterants to whiten and increase the weight of their merchandise. Not surprisingly, the bakers did not feel very kindly toward him, and they disrupted some of his lectures with episodes of mob violence.[4]

The bakers were not Graham's only enemies. Physicians also attacked Graham's views, especially his advocacy of a vegetarian diet, on the traditional medical grounds that moderation was the best policy in diet as in everything else. Over the course of the next century the American medical establishment was forced to confront a number of health lecturers and writers influenced by Grahamist ideology. Besides promoting vegetarianism and the consumption of whole-wheat bread, these dietary reformers were highly critical of orthodox medicine, chiding doctors for seeking to cure the symptoms of a disease (often using dangerous and ineffective therapies) instead of correcting the poor eating habits and other unhygienic practices that caused the illness. In turn, most regular physicians viewed the dietary reform movement as just another quackish medical cult that needed to be wiped out. Because whole-wheat bread was so closely associated with this movement, the nation's doctors tended to view with suspicion anyone who endorsed the consumption of this food.[5]

In the meantime the infant science of nutrition began to produce facts that could be used to assess the healthfulness of bread. In the mid-nineteenth century scientists performed chemical analyses on both white and whole-wheat flour, and determined that whole-wheat had the higher protein, fat, and mineral content. This discovery led some of them to recommend the consumption of the darker flours. But a new line of research begun later in the century converted many nutrition scientists into white-flour advocates. When American and German chemists conducted balance experiments to measure the input and output of the chemical components in food, they found that white flour was the more digestible food. Because some of the nutritious matter found in whole-wheat flour clung to the bran that passed undigested through the human body, people actually obtained more calories and protein from a loaf of white bread than from a loaf of whole-wheat.[6]

Nutritionists' views on bread changed once and for all beginning in the 1910s. During that decade nutrition scientists discovered the presence of hitherto unrecognized substances in food that were necessary for the growth and well-being of animals and humans. They found that one of these substances, which became known as vitamin B, was present in the germ and bran of wheat and other grains. (Over the course of the next two decades it was determined that vitamin B was actually a complex of several vitamins.) Researchers also found that the milling technology introduced into the United States in the 1870s produced a 70 to 72 percent–extraction white flour that was nearly devoid of this vitamin. Although there was no hard evidence that Americans suffered from ill health as a result of eating bread made from this flour, researchers determined that the devastating outbreak of beriberi that had been plaguing much of eastern Asia since the 1880s was caused by the consumption of highly milled white rice. (The thiamine-deficiency disease beriberi is an illness that mainly produces neurological symptoms, which untreated can lead to paralysis and death.) Furthermore, feeding experiments conducted during the 1910s and 1920s confirmed that rats and other laboratory animals subsisting on a diet consisting mainly of whole-wheat bread fared much better than those subsisting on a white bread diet. As a result, nutritionists began to urge

Americans to eat whole-wheat and high-extraction wheat breads instead of white bread, even though they assumed that this substitution was not necessary if the rest of the diet contained a sufficient amount of vitamin- and mineral-rich fruits, vegetables, and dairy products.[7]

Despite this recommendation, the medical establishment continued to endorse the use of white bread throughout the interwar period. Nutrition- ists' support for whole-grain foods was based largely upon the results of animal feeding experiments, but physicians were generally unimpressed by vague notions of positive health derived from research on lower forms of life. Doctors were primarily responsible for treating illness in humans, and there was no real evidence that vitamin deficiency diseases were prevalent among adults in the United States. In fact, the one illness associated with the consumption of white flour, beriberi, was quite rare in this country. (Doctors did suspect that poor nutrition was somehow responsible for pellagra, a disease frequently seen among poor Southerners who subsisted on a diet of corn meal, salt pork, and molasses, but it was not known that this disease was caused by a deficiency of niacin until 1938.) As a result, many doctors dismissed concerns about low intakes of these nutrients. The notion that minor shortfalls of vitamins and minerals might cause ill- ness without producing the full-blown deficiency diseases was slow to influence the profession's thinking. Because there was so little understand- ing of the physiological role played by these nutrients, physicians were unable to recognize signs of incipient deficiency disease. Furthermore, there was almost no quantitative knowledge about human nutritional requirements, which was needed for doctors to make accurate assessments of the adequacy of their patients' diets. Finally, a significant segment of the medical profession worried excessively about the harmful effects of too much fiber in the diet.[8] Consequently, most physicians saw little reason to recommend the increased consumption of whole-grain foods. Instead, they adopted a negative stance toward the use of whole-wheat bread, espe- cially since they associated this product with their longtime enemies.

Yet even as the *Journal of the American Medical Association (JAMA)* issued one last editorial in defense of white flour and bread in April 1939,[9] the premise that the consumption of these foods posed no health hazard to Americans had been seriously undermined. In the mid to late 1930s, after years of trying, chemists succeeded in isolating, identifying, and syn- thesizing a number of vitamins, including thiamine, riboflavin, and nia- cin. Thanks to this development, researchers began to acquire a better understanding of the role these substances played in the human body, and to develop methods to detect minor nutritional deficiencies. Clinicians using these new diagnostic techniques found that many Americans con- sumed less than adequate levels of several vitamins. In addition, as the economic hard times dragged on, international, national, and local gov- ernmental bodies undertook investigations of the impact of the Depres- sion on human health. Dietary surveys conducted under their auspices provided more information on what foods Americans were actually eating. These groups also formulated quantitative requirements for vitamins and minerals. By comparing the survey data to the new dietary standards, researchers were able to generate actual statistics on the nutritional status of the population. They found that a large percentage of Americans were undernourished, and that many more were living on the brink of danger. In contrast to the results obtained from animal feeding studies, this new evidence proved convincing to physicians. As a result, by the end of the 1930s most health authorities had come to accept the view that millions of Americans were suffering the ill consequences of eating too many foods containing too few vitamins and minerals. In recognition of this problem, the American Medical Association (AMA) announced in August 1939 that it considered the restoration of vitamins and minerals to white flour and bread to be a prudent public health measure.[10]

As the country began to mobilize for war in 1940, government officials began to worry that poor eating jeopardized the nation's ability to defend itself, especially after it was revealed that 20 percent of the first one million American men drafted had been rejected as unfit for duty, and that according to the deputy director of the Selective Service System, Brigadier General Lewis B. Hershey, "perhaps one-third of the rejections were due either directly or indirectly to nutritional deficiencies."[11] Most disconcerting to authorities was the lack of thiamine in the American diet, because a deficiency of this vitamin was believed to cause fatigue, depression, irritability, and an unwillingness to work.[12] In response, the Roosevelt administration organized a federally coordinated nutrition program to improve the nation's eating habits. In the end, this program was little more than a massive educational campaign involving hundreds of state and local governmental and nongovernmental organizations that attempted to teach Americans, especially women, the rules of good nutrition and the actions that they could take to protect their health and the health of their families. The program's most enduring legacy, however, was its sponsorship of the proposal to enrich white flour and bread with thiamine, riboflavin, niacin, and iron. In January 1941 program leaders announced that an agreement had been reached with the milling and baking industries to begin production of enriched white flour and enriched white bread according to specifications set by the Food and Drug Administration and supported by the AMA.[13]

Enrichment proponents never imagined that the addition of vitamins and minerals to cereal products and other foods would eliminate malnutrition in the United States. They were fully aware, as were their critics, of the link between poverty and malnutrition. Food consumption surveys, like the one conducted by the USDA, had definitively shown that the lower a family's income, the more likely that its diet was deficient.[14] American physicians engaged in the new field of vitamin therapy also knew that the poor were most likely to develop nutritional deficiencies because their diets did not contain enough meat, dairy products, and fresh fruits and vegetables — foods that contained lots of vitamins and minerals but were relatively expensive. Enrichment proponents readily acknowledged that low-income Americans needed more money to spend on food in order to obtain nutritionally optimal diets. As a result, they did not consider food enrichment as a way to sidestep governmental efforts to increase the food purchasing power of the poor, even though it is true that nutritional authorities at the time were generally reluctant to become involved in controversial political matters.[15]

On the other hand, the proponents of cereal enrichment did not believe that the *particular* dietary deficiencies that they were attempting to correct were fundamentally the result of poverty, but of changes to the food supply that had taken place since the introduction of new milling technologies in the late nineteenth century. They maintained that the vitamin and mineral losses suffered by wheat and other grains during the refining process put all Americans, not just the poor, at risk of developing nutritional deficiencies, especially in regard to thiamine and niacin, since cereal grains are the predominant source of these two vitamins in nature. Because thiamine and niacin are present in much smaller amounts in various other foods, the diets of most Americans contained enough of these vitamins to stave off beriberi and pellagra, but enrichment proponents contended that mild deficiencies of these two vitamins (and of riboflavin and iron as well) were quite common in the United States, even though these deficiencies were seldom diagnosed correctly, for very few physicians were trained to recognize them. The situation in regard to thiamine was considered especially bleak, because thiamine is needed by the body to metabolize carbohydrates, and the consumption of white sugar, which is pure carbohydrate, had risen dramatically in the United States since the mid-nineteenth century. In an influential 1938 article, New York University professor of medicine Norman Jolliffe calculated that the American diet contained just one-third as much thiamine as it had one hundred years before, and that people would have to eat an impossible amount of thiamine-containing foods — 625 grams of fruit, 600 grams of potatoes, 880 grams of other vegetables, and 1,260 cubic centimeters of milk — to compensate for the decreased intake of this vitamin resulting from the consumption of so much white flour and white sugar in the modern diet.[16]

Given this understanding of the situation, there were just three ways to deal with the problem caused by the nutritional losses that occurred when grains were milled. One was enrichment. The second method — supplying Americans with the deficient nutrients via pills and capsules — was rejected by just about everyone because of the relative costliness of this approach, the difficulties involved in getting people to take their vitamins, and most of all, a philosophical opposition to using nonfood items to meet the normal nutritional needs of individuals.[17] The third approach involved teaching Americans to replace the refined grains in their diets with nutritionally superior *natural* foods: either whole or lightly milled grain products or breads made with vitamin- and mineral-rich ingredients, such as yeast, wheat germ, soy flour, and milk. The latter method was the one generally favored by nutritionists and especially by those who criticized enrichment.

Nutritionists' preference for natural foods stemmed in large part from a hard lesson that the profession had learned more than twenty years before. Just before and after the turn of the twentieth century, prior to the discovery of vitamins, nutritionists believed that an adequate diet was simply one that contained a certain number of calories and a certain amount of protein. Hoping to prove the utility of their profession in order to win government funding, leading nutritionists offered advice on how the poor could improve the quality of their diets without having to spend more money on food. They urged the poor to purchase foods that offered the most calories and grams of proteins per dollar and stop wasting their money on costly cuts of meat and such needless extravagances as fresh fruits and vegetables. Had anyone actually followed this advice, of course, the country would have witnessed an outbreak of scurvy and several other deficiency diseases. Then, starting in the 1910s, a revolution took place in the field of nutrition, as researchers discovered the existence of vitamins and developed an appreciation of the nutritional significance of minerals and amino acids. By the end of the 1920s, researchers had learned that there were over thirty essential nutrients, that these nutrients were differentially distributed in a wide variety of foods, and that some nutrients were easily susceptible to destruction during the processing, transportation, and storage of food. As a result, nutritionists now began to recommend that the key to good health was a varied diet of fresh *natural* foods.[18]

The opposition of some nutritionists to the cereal enrichment program in the late 1930s and the 1940s was largely based upon their profession's long-standing preference for natural foods. The three vitamins and one mineral being added to enriched flour were not the only nutrients destroyed during the milling process, but health authorities did not have convincing evidence that these other substances were lacking in the American diet, so they were not included in the enrichment formula. The critics of enrichment, however, did not see the logic or the wisdom in removing valuable nutrients from a natural food, and then adding only some of them back at an increased cost to consumers, no less. They worried about possible deficiencies of the unrestored substances, including some whose nutritional value or even whose existence might not yet be known. Having been embarrassed by their predecessors' unsound dietary recommendations in the pre-vitamin past, they did not want to make the same mistake again. They also pointed to recent evidence that indicated that consuming large amounts of some vitamins without a corresponding increase in the intake of other vitamins could produce symptoms of disease. Thus the opinion of Anton Carlson, the well-known professor of physiology at the

University of Chicago: "Human dietary safety on this front would seem to be to go back to first principles, *putting the whole grain into the flour and the bread.* … On the whole we can trust nature as to the genuine nutritive elements in the whole grain; yes, trust nature further than the chemist and his synthetic vitamins."[19]

Enrichment proponents were actually quite sympathetic to the position taken by their critics. This was especially true of the two leaders of the enrichment campaign: Russell M. Wilder, head of the Department of Medicine at the Mayo Foundation and first chairman of the National Research Council's Committee on Food and Nutrition, and Robert R. Williams, the chemist who first synthesized thiamine. Although Wilder and Williams were not particularly worried about the consequences of restoring just three vitamins and one mineral to enriched flour, they conceded that whole-wheat flour was a superior product, and they reassured the critics that the formula for enriched flour would be modified if evidence emerged to show that a change was desirable. They also acknowledged the superiority of natural foods by recommending that the amount of thiamine, niacin, and iron in enriched flour be equivalent to the amount of these nutrients in whole-wheat flour. In addition, they successfully opposed a proposal to make calcium a mandatory ingredient in enriched flour, even though dietary surveys showed that the American diet was deficient in calcium, on the grounds that wheat did not naturally contain a significant amount of this mineral. "To allow one food to be fortified with excess quantities of vitamins or minerals and with vitamins or minerals which are not indigenous to that type or class of food is to invite multiple corrections of the diet which may add up to a decided overcorrection," Williams cautioned. "Nature knows far more than we do."[20]

Yet even though enrichment proponents shared their critics' preference for natural foods, they did not agree with them that teaching Americans to select whole and lightly milled grain foods in place of refined grains was the best way to increase intakes of the deficient nutrients. They felt that nutrition education was a slow process at best, and that the lower-income families who most needed the extra vitamins and minerals were the least likely to change how they ate. Cultural anthropologists had been saying in recent years that it was both difficult and risky to alter people's food choices, since food habits were deeply ingrained within the culture and often provided socially valuable functions.[21] The preference of the poor and the working class for white bread was considered especially impervious to change because of the long-standing status connotations held by this food. Since Roman times the consumption of white bread had been considered a mark of affluence and high status in the West due to the sweet taste, fine texture, pure color, excellent baking qualities, and high cost of white wheaten flour in relation to other flours. Once the lower classes of London, Paris, and other European cities could afford, by the eighteenth century, to purchase white bread on a regular basis, they had shown a great reluctance to give up eating this food whenever hard times and stingy governments forced them to do so.[22] Although enrichment proponents were aware that certain ethnic groups, such as the Russians and Scandinavians, enjoyed eating breads made from darker flours, they doubted that very many Americans could be persuaded to eat high-extraction wheat breads since, as Robert Williams and Russell Wilder argued, these foods had never regained popularity in any place where white bread had become the primary staple food. Enrichment proponents constantly reminded their critics that just 2 percent of all flour sold in the United States was whole-wheat, and that most of it was eaten by members of the middle class, who valued this food for its health-promoting qualities, but who, unlike the poor, did not depend much on bread for sustenance. The chief advantage of enrichment, according to its promoters, was that nobody's food habits really had to change, since enriched flour looked and tasted exactly like white flour. If just enough people could be convinced to buy enriched flour and bread, the prices of these products would drop, unenriched white flour would be forced off the market, and the poor and everyone else who liked to eat white bread would benefit without really trying. Yale University biochemist George R. Cowgill explained the fundamental premise behind enrichment in a December 1939 position paper for the AMA as follows:

> As a free people must necessarily exercise choices when securing their foods, one basic objective of all efforts directed toward improvement of our dietaries should be to see that, with the various foods readily available, it is exceedingly difficult for any one over a reasonable period, whether he knows a great deal about the science of nutrition or not, to make a set of food selections that is seriously in error. Putting the matter another way, we should aim at making the conditions of food supply such as to yield with the greatest of ease diets that are excellent in every respect.[23]

Critics of enrichment, of course, did not consider a diet containing enriched flour to be "excellent in every respect." They challenged the claim that people could not be taught to like whole-wheat bread or equally nutritious breads made with vitamin- and mineral-rich natural ingredients. "[I am] still optimistic enough to believe that it is well worth while to try to popularize the best bread which we can make," declared Johns Hopkins University biochemist Elmer V. McCollum, one of the codiscoverers of vitamin A and the country's most prominent nutrition scientist. Critics including McCollum contended that the only reason Americans were not eating much whole-wheat bread was simply because no serious attempt had ever been made to encourage them to do so. (Lurking behind this statement was the belief that the milling industry had never really tried very hard to increase demand for whole-wheat flour because the industry profited from selling the bran and the germ as animal feed.) University of California-Berkeley nutritionist Agnes Fay Morgan pointed to her own health-conscious state as proof that people could learn to like whole-wheat bread, for nearly one-third of all bread sold in California was of the wheat or whole-wheat variety.[24]

The two chief proponents of enrichment, Russell Wilder and Robert Williams, remained unconvinced by such optimistic assessments of the power of education. They pointed out that on several occasions since the 1910s millers, bakers, and nutritionists had tried to promote the widespread use of whole-wheat or high-extraction wheat breads, but in every case the attempt had failed, no matter how much money and effort had been spent on publicity. They considered the two most recent failures to be especially illustrative. In 1937 the Swiss government had tried to encourage its citizens to eat a high-extraction wheat bread by taxing the by-products of white flour milling, which caused the price of white bread to be 20 percent above that of the darker bread. At first the plan seemed to work: the wheat bread accounted for 75 percent of all bread sales. By the end of 1938, however, the figure had dropped to just 11 percent, and by 1944 the share of the market held by the wheat bread was exactly the same as that held by nonwhite breads before the plan was implemented. In 1941, America's largest bread maker, Continental Baking Company, sought to take advantage of the national nutrition program by introducing with much fanfare a new high-extraction wheat bread called Staff Bread. After an initial burst of sales, consumer interest in this product declined rapidly, and the company was forced within a year to abandon its attempt to market the bread nationally. According to Wilder and Williams, the rapid failures of these two campaigns, both of which began with lots of publicity and great success, proved what everyone already knew: that most people really did not like the taste, appearance, and baking qualities of whole-wheat flour. As for the data from California suggesting that consumers could learn to enjoy wheat bread, Wilder and Williams dismissed these statistics as meaningless on the grounds that most wheat breads sold in this country were nutritionally inferior to enriched bread because they were actually made from mixtures of whole-wheat flour and white flour, with the former seldom exceeding one-third of the total composition of the flour. The way Wilder and Williams saw the matter, the critics of enrichment were just a bunch of impractical elitists, who were undermining the program: "Would you rather have a selected few benefit by whole wheat teaching or give 'enriched' products to the masses?" Williams asked an audience of skeptical home economists.[25]

The enrichment controversy was not confined to the United States, but the outcome was different elsewhere. In July 1940 the British government, which had more reason than American officials to be concerned about the effects of the war on the health of its citizens, made a surprise announcement that it had reached an agreement with the milling industry to have all white flour produced in the country enriched with thiamine. (It also announced its intention to add calcium to enriched flour at a later date, but it did not plan to add riboflavin, niacin, or iron, as in the United States.) Although the nation's two leading medical journals initially expressed cautious support for the action, British nutritionists condemned the new flour as a woefully inadequate product, and they began urging the government to mandate the use of a high-extraction flour. Thanks to the vocal support of several members of Parliament, the foremost of whom belonged to the Food Education Society, an organization descended from the Grahamist tradition, nutritionists convinced the government in December 1940 to market an 85 percent–extraction wheat bread that sold for the same price as enriched white bread. This halfway measure, however, did not satisfy the anti-enrichment party, especially when it learned that the National Wheatmeal loaf, as the new product was called, never accounted for more than 9 percent of total bread sales. Following more protests, the British government finally decided in March 1942 to halt the production of enriched flour and to fix the extraction rate at 85 percent. Health considerations, however, had less to do with this decision than the need to reduce wheat consumption in order to conserve shipping space. Not only is less wheat needed to produce a loaf of whole-wheat bread than to produce a loaf of white bread, but it was expected that people would eat more potatoes and less bread if they were forced to eat a bread containing more bran. (When more shipping space became available in late 1944, the government lowered the extraction rate to 80 percent.) During this time of crisis and self-sacrifice, the British people accepted the loss of their white bread without complaint.[26]

When the war started, nutritionists in Canada began to agitate for the improvement of white flour. Like their colleagues in other countries, they too were opposed to enrichment, so they worked with their nation's milling industry to find a way to alter its manufacturing process so as to produce a new kind of white flour that contained a greater amount of the natural vitamin and mineral content of wheat than was present in regular white flour. Because of the relatively high natural thiamine content of the hard wheat grown in that country, the new "Canada Approved" flour contained enough thiamine to metabolize all the carbohydrates in bread, although it contained less of this vitamin than was found in American enriched flour. The use of this food, however, was not mandatory, and its market share in 1944 ranged somewhere between 7 and 40 percent. Meanwhile, in accordance with the tenor of their country's pure food law, Canadian officials outlawed the addition of synthetic vitamins to flour and bread as an adulteration.[27]

The story in Newfoundland, which did not become part of Canada until 1949, was completely different. The people of that country were extremely poor, and the vitamin deficiency diseases xerophthalmia (vitamin A), scurvy (vitamin C), and most notably, beriberi (this term was probably used to identify a combination of vitamin B-complex deficiencies) were rife there in the first half of the twentieth century. Although few Newfoundlanders actually died as a direct result of these illnesses, the morbidity and infant mortality rates were very high, and people were often incapacitated for several months following a long winter when they had access to few fresh foods and subsisted on a diet consisting mainly of white bread, molasses, and tea. As early as 1916 doctors and public health workers tried to teach Newfoundlanders to eat whole-wheat bread, but this approach had very limited success. When the economic and health situation deteriorated during the 1930s, government officials required that all welfare recipients be provided a high-extraction wheat bread in place of white bread. This action was extremely unpopular with labor unions, industry representatives, and most of all, with the recipients themselves, who protested the policy in the streets and even in song. When World War II came, the economy improved, but the malnutrition problem seemed to become even worse, as people came off the dole and returned to eating white bread. Troubled by the inability of Newfoundlanders to perform their jobs at the Canadian and American military bases located in their country, government officials debated whether to mandate the use of a high-extraction wheat flour or the use of an enriched white flour made according to the American formula. Given their past experience with brown bread, in July 1944 they settled on the latter approach. The action produced a tremendous decline in vitamin deficiencies according to every indicator — dietary surveys, clinical examinations, and morbidity and mortality rates — used to detect these diseases. As a result, when Newfoundland agreed to join Canada, it did so on the condition that it could keep its mandatory enrichment law.[28]

Although circumstances unique to each country accounted for the difference in outcomes, the rejection of cereal enrichment in Britain and Canada was likely due in part to the greater influence of scientists in nutrition policy making than in the United States, where the medical profession played the leading role. Unlike nutritionists, who had a holistic view of the relationship between food and health and always promoted dietary improvement through education, physicians were willing to tackle specific nutritional deficiencies individually and were very comfortable using chemical substances, whether they were synthetic vitamins or drugs, to treat disease. The nature and extent of the malnutrition problem also affected the course of events in each country. Unlike the situation in Newfoundland and the American South, frank deficiencies diseases like beriberi and pellagra were rare in Britain and Canada. As a result, authorities there could afford to wait the time needed to educate people to eat whole-grain foods. In contrast, health officials in the other two countries confronted potentially fatal diseases that needed to be dealt with immediately. As the Newfoundland experience demonstrated, they knew that it was not going to be easy or popular to persuade people to give up white bread. Enrichment, which was a relatively low-cost and easy-to-implement policy, required no real change in eating habits. One additional reason for the adoption of enrichment in the United States was the personal influence of Russell Wilder, a charismatic figure described by one nutritionist as a Greek god.[29]

On the terms set by its proponents, cereal enrichment proved to be a great success. Per capita consumption of thiamine, riboflavin, niacin, and iron increased between 35 and 42 percent between the late 1930s and 1945. Consumption of enriched grain products accounted for between 42 and 72 percent of this increase; the rest was due to the general improvement in the American diet that occurred during the prosperous war years. Intakes of the four nutrients are even higher today thanks to the voluntary enrichment of ready-to-eat breakfast cereals (which began in 1955) and slight upward revisions in the enrichment formula that were made in the 1970s and early 1980s. It is more difficult to prove that enrichment by itself was responsible for an actual improvement in the health of Americans, but striking declines in the incidence of the vitamin deficiency diseases that occurred within a decade following the adoption of enrichment are highly suggestive. In the late 1940s and early 1950s urban hospitals that had previously treated many patients suffering from these illnesses reported that they were having difficulty finding cases to show their interns and medical students. Prior to enrichment, chronic alcoholics had been extremely likely to develop these diseases, but in 1948 and 1949 no more than 14 out of 16,000 alcoholic inmates at the Chicago House of Corrections were diagnosed as having them. Undoubtedly the best evidence of the success of enrichment has been the almost total eradication of pellagra in the United States. In 1938, the year when voluntary cereal enrichment began, 3,200 people died from pellagra; by the time the first mandatory enrichment law was passed four years later, the number had been cut in half; by 1950 the number had dropped to less than three hundred. Over the same period the decline in the morbidity rate was even greater.[30] Although some have argued that wartime and postwar prosperity, which enabled poor Southerners to obtain better diets, was chiefly responsible for the conquest of pellagra, both the timing and the extent of the decreases in mortality and morbidity rates suggest that cereal enrichment was the primary cause.[31]

Because of the perceived effectiveness of enrichment as a public health measure, in 1996 the Food and Drug Administration added another nutrient to the formula for enriched flour and bread. By the early 1990s sufficient evidence had been gathered to convince American medical authorities that low intakes of the vitamin folic acid by pregnant women were responsible for some of the approximately four thousand pregnancies affected each year by neural tube birth defects, such as spina bifida. After some debate, public health officials decided that adding 140 micrograms of folic acid to 100 grams of enriched flour was the most effective and safest way to deal with this problem. Early evidence has already indicated that this action has been successful in reducing the number of neural tube birth defects in the United States.[32]

Despite these achievements, nutritionists today continue to bad-mouth white flour, and to urge Americans to eat more whole-grain foods. In the past decade or so, research has begun to confirm what the critics of enrichment had predicted in the 1940s: that substances removed from whole grains during the refining process but not restored to enriched foods possess health value. Folic acid is one example of this, although authorities decided that enriched flour should contain twice the naturally occurring amount of this vitamin in order to be effective in preventing birth defects. Possessing even greater potential health significance are recent studies that have linked the consumption of whole grains to reduced risks for heart disease and several types of cancer. Although the exact mechanisms responsible for these benefits are not fully known, researchers suspect that a number of vitamins and minerals and previously unappreciated non-nutrients, such as fiber and phytochemicals, play key roles. White flour, meanwhile, is disparaged not only for its chemical shortcomings but also for its highly refined physical structure, the combination of which is believed to elevate glucose levels and insulin demand, which in turn can lead to diabetes and obesity.[33] In general, the preference of nutritionists for a varied diet of fresh natural foods remains today as strong as ever, and as a result they are still less than enthusiastic about improving the quality of the food supply by fortifying processed foods with synthetic vitamins and minerals. On the other hand, nutritionists have never called for an end to cereal enrichment; even Columbia University professor of nutrition and education Joan Dye Gussow, a vocal critic of the fortification of processed foods, has acknowledged that cereal enrichment was a sound idea.[34] It should also be remarked that because of the popular demand for white bread, both Britain and Canada officially abandoned their opposition to enrichment back in 1953.[35]

In assessing the effectiveness of the enrichment program, it should be remembered that enrichment was never meant to be a nutritional panacea. The program was a technological fix designed to correct a mistake that technology itself had caused, targeted at a few specific deficiencies that were causing known health problems. Furthermore, if our current obesity problem has taught us anything, it is the relative ineffectiveness of education in changing food habits when many competing forces are operating against success. As recent studies have shown, despite recommendations that people eat three servings of whole-grain foods each day, the average American eats only one, and the lower one's income and the less health conscious one is, the less likely it is that one consumes these foods.[36] It thus seems that enrichment advocates were wise to propose an approach whereby Americans, no matter how much they know about nutrition, would immediately get the vitamins and minerals they most needed without having to make radical changes to their diets. At least then nutritionists would have the luxury of time to teach people to eat more whole-grain foods. To paraphrase Robert Williams, when people are starving, nine-tenths of a loaf for everyone is clearly better than a whole loaf for a relative few.

Notes

1. Hazel K. Stiebeling, *Are We Well Fed?: A Report on the Diets of Families in the United States*, USDA Miscellaneous Publication No. 430 (Washington: USDA, 1941).
2. Thomas Parran, "Health and Medical Preparedness," *JAMA* 115 (6 July 1940): 51; Federal Security Agency, Office of the Director of Defense Health and Welfare Services, *Proceedings of the National Nutrition Conference for Defense* (Washington: Government Printing Office (GPO), 1942).
3. Russell M. Wilder and Robert R. Williams, *Enrichment of Flour and Bread: A History of the Movement*, Bulletin of the National Research Council No. 110 (Washington: NRC, 1944); Youngmee K. Park, et al., "History of Cereal-Grain Product Fortification in the United States," *Nutrition Today* (May/June 2001): 124–37.
4. Stephen Nissenbaum, *Sex, Diet, and Debility in Jacksonian America: Sylvester Graham and Health Reform* (Westport, CT: Greenwood Press, 1980), 5–15, 20–21, 86–101; James C. Whorton, *Crusaders for Fitness: The History of American Health Reformers* (Princeton, NJ: Princeton University Press, 1982), 38–49, 57; William B. Walker, "The Health Reform Movement in the United States, 1830–1870" (Ph. D. diss., Johns Hopkins University, Baltimore, MD, 1955), 57–61; R. A. McCance and E. M. Widdowson, *Breads, White and Brown: Their Place in Thought and Social History* (London: Pitman Medical Publishing, 1956), 48–49.
5. Whorton, *Crusaders for Fitness*, 73–76, 109–11, 127–31; Walker, 42–43, 90–92; Regina Markell Morantz, "Nineteenth Century Health Reform and Women: A Program of Self-Help," in *Medicine without Doctors: Home Health Care in American History*, eds. Guenter B. Risse, Ronald L. Numbers, and Judith Walzer Leavitt (New York: Science History Publications, 1977), 74–77.
6. McCance and Widdowson, 53–54.
7. The analysis offered in this and the next two paragraphs is based on material from chapters 4 and 5 of the author's forthcoming Ph.D. dissertation, preliminarily titled "Food Faddism and Nutrition Science in the United States, 1910–1965."
8. James C. Whorton, *Inner Hygiene: Constipation and the Pursuit of Health in Modern Society* (New York: Oxford University Press, 2000), 218–20.
9. "The Nutritive Value of Wheat Flour and Bread," *JAMA* 112 (15 April 1939): 1461–62.
10. AMA Council on Foods, "Annual Meeting of the Council on Foods," *JAMA* 113 (19 August 1939): 681.
11. Lewis B. Hershey, "Selective Service and Its Relation to Nutrition," in *Proceedings of the National Nutrition Conference for Defense*, 63–67. It should be noted that the figure of one-third of all rejections being due to malnutrition was based on the controversial assumption that tooth decay was caused by poor nutrition. Bad teeth accounted for nearly 20 percent of all rejections. See Harvey Levenstein, *Paradox of Plenty: A Social History of Eating in Modern America*, (New York: Oxford University Press, 1993), 65.
12. Ray D. Williams et al., "Observations on Induced Thiamine (Vitamin B_1) Deficiency in Man," *Archives of Internal Medicine* 66 (October 1940): 785–99; "Vitamins for War," *JAMA* 115 (5 October 1940): 1198–99.
13. "Will Offer Bread Vitamin-Enriched," *New York Times*, 30 January 1941, 14; "Nutritionally Improved or Enriched Flour and Bread," *JAMA* 116 (28 June 1941): 2849–53.
14. Hazel K. Stiebeling and Esther F. Phipard, *Diets of Families of Employed Wage Earners and Clerical Workers in Cities*, USDA Circular No. 507 (Washington: GPO, 1939).

15. In a 1995 article, historian Anne Hardy contrasted the attitude of American enrichment proponent Robert R. Williams with that of the British nutritionists who led the United Nations's international campaign against beriberi after World War II. According to Hardy, the British experts opposed cereal enrichment because they perceived the problem of malnutrition as essentially one of economics, and they thus promoted structural and educational programs that worked to increase consumption of whole and lightly milled grain products and other foods rich in vitamins and minerals. Williams, Hardy implied, saw beriberi as an isolated deficiency disease that could be eradicated with a technological fix. See Anne Hardy, "Beriberi, Vitamin B_1 and World Food Policy, 1925–1970," *Medical History* 39 (1995): 61–77. I disagree with this interpretation. Enrichment proponents, including Williams, believed that enrichment was simply the cheapest and most effective way to help the poor obtain those nutrients destroyed in the milling process. Furthermore, they saw no conflict between using technologies, like enrichment, to combat malnutrition and the enactment of programs aimed at improving living standards. It is much more likely that the opposition of British nutritionists to enrichment was due to their profession's long-standing preference for whole-grain foods than to a greater appreciation of the structural causes of malnutrition. See Wilder and Williams, *Enrichment of Flour and Bread*, 79–81.

16. Wilder and Williams, *Enrichment of Flour and Bread*, 63–68; Robert R. Williams and Tom D. Spies, *Vitamin B1 (Thiamin) and Its Use in Medicine* (New York: Macmillan Co., 1938), 84–86, 106–110; R. R. Williams, "Our Vitamin B_1 Supply in Relation to Human Needs," *Bulletin of the New York Academy of Medicine* 14 (October 1938): 641–46; Norman Jolliffe, "A Clinical Evaluation of the Adequacy of Vitamin B_1 in the American Diet," *International Clinics* n.s. 4 (1938): 46–66. Sugar consumption increased from about 10 pounds annually per person in 1821 to about 85 pounds around 1940, by which time it furnished 13 to 17 percent of all calories in the average American diet.

17. *Proceedings of the National Nutrition Conference for Defense*, 105, 110–11.

18. Material for this paragraph comes from Chapter 1 of the author's forthcoming dissertation. Also see Harvey A. Levenstein, *Revolution at the Table: The Transformation of the American Diet* (New York: Oxford University Press, 1988), 44–59, 72–75; and Hamilton Cravens, "Establishing the Science of Nutrition at the USDA: Ellen Swallow Richards and Her Allies," *Agricultural History* 64 (spring 1990): 122–33.

19. Agnes Fay Morgan, "Fortification of Foods with Vitamins and Minerals: The Basic Nutrition Principles," *Milbank Memorial Fund Quarterly* 17 (July 1939): 221–29; Lydia J. Roberts, "Fortification in a General Program for Better Nutrition," in *Milbank Memorial Fund Quarterly* 17 (July 1939): 236–39; A. J. Carlson, "What Is Wrong with the American Diet?," *Northwest Medicine* 41 (October 1941): 335–36; Samuel Lepkovsky, "The Bread Problem in War and in Peace," *Physiological Reviews* 24 (April 1944): 266–73.

20. Wilder and Williams, *Enrichment of Flour and Bread*, 10–11, 22–28; R. R. Williams, "Fortification and Restoration of Processed Foods," *Industrial and Engineering Chemistry* 33 (June 1941): 718–19; Russell M. Wilder, "Vitamin Restoration of Foods as Viewed by the Physician," *Scientific Monthly* (October 1941): 295–96, 302; "In the Matter of Definitions and Standards of Enriched Flours, etc.," Docket No. FDC–21(a), April 1943, pp. 273–77, 325–27, 405–412, 493–516, Public Docket's Office, Food and Drug Administration, Rockville, MD.

21. Levenstein, *Paradox of Plenty*, 72–73; Rebecca L. Spang, "The Cultural Habits of a Food Committee," *Food and Foodways* 2 (1988): 374–75.

22. McCance and Widdowson, 36–42; Christian Petersen, *Bread and the British Economy, c1770–1870*, ed. Andrew Jenkins (Aldershot, Hants, England: Scolar Press, 1995), 32–36; Steven Laurence Kaplan, *The Bakers of Paris and the Bread Question, 1700–1775* (Durham, NC: Duke University Press, 1996), 30–36.

23. Wilder and Williams, *Enrichment of Flour and Bread*, 78; Robert R. Williams and Russell M. Wilder, "Why Enriched Bread?" *Journal of the American Dietetic Association* 18 (April 1942): 225–26; Russell M. Wilder, "The Case for Enriched Flour and Bread," attachment to letter from Wilder to John B. Youmans, Vanderbilt University, 8 December 1941, 1–2, 10–12, 19–21, in Box 16 (NRC, 1941), John B. Youmans Papers, National Library of Medicine, Bethesda, MD; Robert R. Williams, "Refined Foods and Nutritional Deficiency," *Medical Woman's Journal* 49 (April 1942): 3–8; George R. Cowgill, "The Need for the Addition of Vitamin B_1 to Staple American Foods," *JAMA* 113 (9 December 1939): 2150.

24. Lepkovsky, 263–66; E. V. McCollum, "Bread 'Enrichment,'" *Maryland Health Bulletin* 17 (March 1945): 2–4; Agnes Fay Morgan, "Education the Key," *Journal of Home Economics* 37 (September 1945): 400–401.

25. Wilder and Williams, *Enrichment of Flour and Bread*, 47, 76–77; Robert R. Williams, "Bread 'Enrichment,'" letter in *Science* 102 (17 August 1945): 180–81; Williams, "Refined Foods and Nutritional Deficiency," 6–7.

26. McCance and Widdowson, 85–100, 112–19; Lepkovsky, 250–60.

27. McCance and Widdowson, 105–106; Lepkovsky, 260–1; Wilder and Williams, *Enrichment of Flour and Bread*, 79–81.

28. James Overton, "Brown Flour and Beriberi: The Politics of Dietary and Health Reform in Newfoundland in the First Half of the Twentieth Century," *Newfoundland Studies* 14 (1998): 1–27; R. M. Wilder, "Enriched Food Products Are Credited for Better Health in Newfoundland," *Food Industries* (July 1949): 44–47.

29. Robert R. Williams, *Toward the Conquest of Beriberi* (Cambridge, MA: Harvard University Press, 1961), 178; Franklin C. Bing, interview by Marie Balsley, 24 October 1980, transcript of tape recording, in Box 9d (Student Notebooks and Curriculum Vitae), Folder 14, Franklin C. Bing Papers, Eskind Biomedical Library, Vanderbilt University Medical Center, Nashville, TN.

30. The increases in intake of thiamine, riboflavin, niacin, and iron were 42 percent, 35 percent, 38 percent, and 35 percent, respectively; the percentages of the increase due to enrichment were 72 percent, 42 percent, 58 percent, and 56 percent, respectively. Faith Clark, Berta Friend, and Marguerite C. Burk, *Nutritive Value of the Per Capita Food Supply, 1909–45*, USDA Misc. Pub. No. 616 (Washington, 1947), table 6; Russell M. Wilder, "Nutritional Health of Adults," in *Proceedings of the National Food and Nutrition Institute*, USDA Agricultural Handbook No. 56 (Washington: GPO, 1953), 52–54; Paul A. Lachance and J. Christopher Bauernfeind, "Concepts and Practices of Nutrifying Foods," in *Nutrient Additions to Food: Nutritional, Technological and Regulatory Aspects*, eds. Paul A. Lachance and J. Christopher Bauernfeind (Trumbull, CT: Food and Nutrition Press, 1991), 73–75.

31. Just three years ago a team of researchers confirmed this conclusion using statistical analysis to evaluate various possible explanations for the virtual disappearance of pellagra in the United States. See Youngmee K. Park et al., "Effectiveness of Food Fortification in the United States: The Case of Pellagra," *American Journal of Public Health* 90 (May 2000): 727–38.

32. Department of Health and Human Services, Food and Drug Administration, "Food Standards: Amendment of Standards of Identity for Enriched Grain Products to Require Addition of Folic Acid," *Federal Register* 61 (5 March 1996): 8781–97; Margaret A. Honein et al., "Impact of Folic Acid Fortification of the U.S. Food Supply on the Occurrence of Neural Tube Defects," *JAMA* 285 (20 June 2001): 2981–86; T. J. Mathews, Margaret A. Honein, and J. David Erickson, "Spina Bifida and Anencephaly Prevalence — United States, 1991–2001," *Morbidity and Mortality Weekly Report* 51 (13 September 2002): 9–11.

33. Walter C. Willett, "The Dietary Pyramid: Does the Foundation Need Repair?" *American Journal of Clinical Nutrition* 68 (1998): 218–19; Joanne Slavin, "Why Whole Grains Are Protective: Biological Mechanisms," *Proceedings of the Nutrition Society* 62 (2003): 129–34.

34. Katherine L. Tucker, "Eat a Variety of Healthful Foods: Old Advice with New Support," *Nutrition Reviews* 59 (May 2001): 156–58; "Position of the American Dietetic Association: Food Fortification and Dietary Supplements," *Journal of the American Dietetic Association* 101 (January 2001): 115–25; Joan Dye Gussow and Sharon Akabas, "Are We Really Fixing Up the Food Supply?" *Journal of the American Dietetic Association* 93 (November 1993): 1300–304.

35. J. C. Bauernfeind and E. DeRitter, "Foods Considered for Nutrient Addition: Cereal Grain Products," in *Nutrient Additions to Food*, 162.

36. Linda E. Cleveland et al., "Dietary Intake of Whole Grains," *Journal of the American College of Nutrition* 19, no. 3 (2000): 331S–338S; Judi F. Adams and Alta Engstrom, "Helping Consumers Achieve Recommended Intakes of Whole Grain Foods," in *Journal of the American College of Nutrition* 19, no. 3 (2000): 339S–344S.

5

Long-Haul Trucking and the Technopolitics of Industrial Agriculture, 1945–1975

SHANE HAMILTON

In 1954 historian David Potter called Americans the "People of Plenty." The exceptional nature of America's economic abundance, argued Potter, has led to a series of exceptional American cultural and political behaviors. One of Potter's key assertions was that American-style democracy has often encouraged the use of technological means to fix complex political problems. One salient example of this, argued Potter, was the mid-twentieth-century introduction of long-haul trucks "moving over a network of national highways [to end] the natural monopoly ...of railroads."[1] In the 1920s and 1930s, truckers began offering railroads serious competition. After a frenzy of federal and state highway building in the 1940s and 1950s made reliable long-distance hauling feasible, trucks came to replace railroads as the primary mode for hauling America's freight by the early 1970s.[2] Relatively few historians have asked why this significant change in transportation modes occurred. Those historians who have, however, tend to agree with Potter's assessment: the problem of regulating monopolistic railroads to insure fair freight rates was just too much political headache for federal administrators. The monopolistic railroads had built up so much political ill will since the days of the Granger antitrust crusades, these historians argue, that by the 1940s Congress decided to use billions of dollars of taxpayer money to build highways, effectively killing the railroad "octopus" by providing free infrastructure to truckers.[3]

93

There are a number of problems with this interpretation of why trucks became the nation's primary haulers of general freight in the latter half of the twentieth century. One especially glaring problem is that most trucking firms, since the Motor Carrier Act of 1935, were regulated until the 1970s as "natural monopolies," just like the railroads.[4] Even more problematic is the assertion that a unified consensus existed among transportation policymakers, who felt that railroads were political-economic monstrosities in need of replacement. A few large railroad corporations proved easier to regulate than thousands of smaller entities; consequently, the Interstate Commerce Commission (the government agency responsible for regulating the railroads in the twentieth century) consistently sided with the largest railroads in legal and political matters.[5] Potter was on to something when he used trucks as an example of a technological fix in an economy of abundance, but antitrust was not the main political problem trucks were meant to solve. The problem of monopoly, of course, particularly in relation to railroads, has been a prevalent issue in American politics since the nineteenth century.[6] I would argue, however, that this was only one among many political issues that made trucking into a politically attractive technological fix, in a variety of institutional settings. This paper focuses on one of the institutions most responsible for encouraging the rapid growth of trucking in the second half of the twentieth century: the U.S. Department of Agriculture (USDA).

Agricultural policymakers, engineers, and economists in the USDA saw trucking in the 1940s and 1950s as a key to solving the decades-old "farm problem." This so-called farm problem was simple in theory: agricultural policymakers in the USDA and in Congress wanted to make sure farmers made enough money to keep them on the farm, while keeping food cheap for urban consumers. Achieving this balance proved extremely difficult in practice. Since the end of the Civil War, science and technology had made American farms incredibly productive, reducing the average price of food for consumers. The unfortunate result for farmers was that they usually grew more food than they could sell at profitable prices. In the agricultural depression of the 1920s and 1930s, agricultural policymakers sought legislative solutions to the farm problem, culminating in the highly controversial Agricultural Adjustment Acts of 1933/1938.[7] The overwhelming productivity of American agriculture after World War II, however, made the New Deal legislative solutions both controversial and obsolete. In the postwar period, as Keynesian ideas took firm hold in the top rungs of economic policymaking while middle-class consumers fretted about the rising cost of living, agricultural policymakers sought to find new, less controversial ways to solve the farm problem. Agricultural abundance, coupled with a supercharged mass consumption economy, shifted the terms of policy debates over the farm problem. Depression-era agricultural policymakers had centered on ways to limit production, but post-World War II agricultural policymakers sought primarily to rationalize the consumption of agricultural products. Technologies for making food distribution cheaper and more reliable were the key focus, and trucking was at the center of this technological fix. Monopoly was not the main political-economic problem that trucking came to solve; instead, trucking was a solution to the political problem of maintaining economic abundance.

Trucking, as agricultural experts in the USDA saw it, would be an integral part of a new kind of agricultural marketing machine — an efficient and flexible system of food distribution. The primary elements of this marketing machine were to be highly mechanized farms, high-tech food processors, and suburban supermarkets. Each of these elements, by practicing economies of scale and by using the latest technologies — from pesticides on farms to forklifts in cold-storage warehouses — could significantly reduce the costs of labor involved in moving food from a farm to a distant consumer. Long-haul tractor-trailers would tie all these elements together, providing a flexible means of quickly and reliably moving relatively small lots of produce, dairy, meat, and grains to food processors and then hauling finished food products directly to supermarkets. The flexibility of trucks — their geographical reach, customized hauling capabilities, and their ability to haul small loads on short notice — were essential components for making this new marketing machine work. The flexibility of trucking allowed supermarkets and food processors to team up to eliminate economic uncertainties (such as seasonal and regional variations in production or potential strikes from unionized workers) from the food distribution chain. In the minds of USDA policymakers, engineers, and economists, this rationalization of food distribution would successfully solve the farm problem once and for all, keeping commodity prices high for farmers while consumer food prices remained steady.

This paper focuses on the transportation of one specific product — frozen foods — to demonstrate the motives and activities of the USDA in helping the postwar food marketing machine develop. After a brief overview of the history of the farm problem in federal politics, I examine the work of a group of engineers at USDA research stations in California who worked on the problem of using tractor-trailers to move frozen foods smoothly from processing plants to supermarkets. The final section of the paper shifts the focus back to Washington, DC, looking at the intertwined work of agricultural economists and policymakers (in the office of the Secretary of Agriculture) in transportation policy debates of the 1950s. Of particular importance was the question of how strictly the Interstate Commerce Commission should regulate truckers hauling frozen foods. By looking at these three groups of USDA agricultural experts (engineers, economists, and policymakers), I hope to give a bird's-eye view of how this governmental institution used technologies and technological expertise to engineer the food sector of the postwar mass-consumption economy.

Most historians who have looked at America's late-twentieth-century mass-consumption economy have generally ignored the importance of technologies in shaping the politics of the period. Social and political historians have examined the importance of labor movements for increasing worker purchasing power;[8] the role of intellectuals, activists, and politicians in creating a republic of "consumer citizens;"[9] and the postwar emergence of macroeconomic policies aimed at achieving economic growth by encouraging mass consumption.[10] This essay, however, is about a less visible, although still significant, form of politics — what historians of technology have called *technopolitics*.[11] This approach seeks not only to understand how political concerns shape technological artifacts, but also how technologies become embedded into *specific* political debates as a means of settling disputes. Trucking became a major economic force in American life in the twentieth century for a wide variety of reasons; the actions of the USDA in encouraging trucking's growth were only part of a myriad of conflicting efforts by various interest groups, private firms, and other governmental agencies. Nonetheless, tractor-trailers became particularly important as political technologies when USDA engineers and policymakers turned to them to redefine the terms of debate in the political economy of postwar agriculture. In this sense, big rigs were not so much a technological fix to the farm problem itself as they were a convenient means for the USDA, as an institution, to refocus its efforts on the creation of a more efficient marketing machine. Furthermore, by using technical expertise in marketing economics and transportation engineering to work on the farm problem, various agents of the USDA helped to push the politics of postwar agriculture out of the public sphere and into the realm of the myteriously technical and administrative. New Deal–era debates about farm policy had engendered large-scale public debates about the question of a just price for commodities; by the late 1960s, agricultural policy debates were essentially focused on the finer details of administratively maintaining economic growth. Like other cases in the twentieth century when policymakers have turned to technical means to solve political problems (nuclear power, smokestack pollution filters, tax policies) the result has been the creation of an increasingly administrative role for the state in matters of political economy.[12]

The Farm Problem

To understand why USDA agricultural experts sought a technological fix to the farm problem, we have to understand just how tricky the problem had become by the 1940s and 1950s. The problem of maintaining a balance between farmers' incomes and consumer food expenses first appeared as a politically salient issue during the Populist movements of the late nineteenth century. Southern farmers reacting to the credit squeeze of the crop lien system, along with Northern Plains farmers struggling with the economic distress of droughts and globalizing wheat markets, called on the federal government to protect farmers from the nation's "money interests" (i.e., landlords, banks, and railroads). Although the Populists failed to elect their presidential candidates in the 1892 and 1896 national elections, they were successful in putting the farm problem on the nation's political agenda.[13] Progressive reformers of the early twentieth century adapted many of the Populists' ideas as new legislation and policies, from the creation of the Interstate Commerce Commission to the establishment of rural producers' cooperatives to improve the leverage of farmers in agricultural markets.[14] These policy efforts had some success in mitigating the farm problem, but even more important were the rising prices for farm products (particularly wheat) that came in the 1910s with expanding global demand. The period leading up to and through World War I thus witnessed a "golden age of agriculture" that significantly defused political agitation by farmers.[15]

The farm problem returned to the nation's political consciousness with a vengeance in the 1920s. Huge surpluses created by production for World War I led to a postwar drop in farm prices and a consequent agricultural depression. Congressmen from rural states reacted by forming a farm bloc devoted to increasing farmer's incomes, either by limiting agricultural production or by guaranteeing farmers a parity price for their crops. Attempts to pass effective legislation like the McNary-Haugen Bill foundered in the 1920s, as farm representatives from different regions of the country could not reach consensus on the proper mechanism for assuring steady farm incomes.[16] When the Great Depression struck in 1929, however, the farm problem became especially acute, as impoverished and desperate farmers called on the federal government for support. Herbert Hoover's Farm Board attempted to implement the least statist proposals discussed in the McNary-Haugen debates (particularly voluntary marketing controls to shore up farm prices) but with little success. Most farmers continued to act as self-interested individuals, refusing to cooperatively reduce their production to increase prices. The agricultural depression continued.[17]

When Franklin Roosevelt came into office, one of his administration's first acts during the famous "First 100 Days" was to put together all of the ideas from the 1920s and Hoover's farm program into the Agricultural Adjustment Act, or AAA. The AAA did eventually shore up some farmer's incomes by creating price supports and production controls, but at the price of forcing thousands of small farmers, tenants, and sharecroppers off their lands.[18] As a consequence, the AAA offended both conservatives who saw it as an affront to free enterprise, and liberals who saw the program as harmful to the least privileged members of rural society. Furthermore, the goals of the AAA were directly at odds with much of the rest of New Deal legislation because raising farm prices only served to increase the cost of food for other members of the New Deal coalition, particularly urban industrial laborers. These aspects of the New Deal farm legislation were controversial, but the programs became especially politically unpopular when Secretary of Agriculture Henry A. Wallace ordered six million baby hogs killed and one-quarter of the southern cotton crop plowed under to increase market prices in 1933 and 1934. Republican critics of the New Deal ridiculed the administration for destroying crops and livestock when millions of Americans were starving and poorly clothed.[19]

Underlying all of these political controversies was the simple fact that American farmers produced too much food in the first half of the twentieth century. Even as the political wing of the U.S. Department of Agriculture was administering legislation to support farm prices and limit the amount of acreage farmers could put into production, its scientific and technological bureaus were successfully encouraging farmers to use pesticides, fertilizers, hybrid crops, and tractors to *increase* their production. Secretary Wallace only very reluctantly ordered the slaughtering of six million baby hogs in 1933, since his fundamental philosophy regarding American agriculture was one of ever-increasing abundance, not limits to production.[20] The USDA's technological and scientific efforts from the late nineteenth century into the 1940s, encouraged by economists such as M. L. Wilson and John D. Black, focused on creating huge, industrial farms where food could be produced factory-style.[21] The AAA made for good headlines and solid political support for the Democrats from large commercial farmers, but the USDA's real efforts to solve the farm problem were, until the post-World War II era, focused primarily on increasing big farmers' production and forcing small, "inefficient" farmers out of the market.[22]

This stance was useful during World War II, when American farmers were called upon to feed the boys overseas. You cannot fight a war on an empty stomach, and America's highly productive farmers profited from the chance to keep the Allied soldiers in prime fighting condition. The devastation of European and Soviet agricultural fields sent demand for American agricultural products soaring. Assured of high prices for their products, American farmers were able to invest heavily in tractors, fertilizers, hybrid seeds, and other technologies for increasing crop yields without concern for overproduction.[23] High demand for American agricultural products pushed prices up, but at the same time, wartime price controls made sure the prices, if not the availability, of food remained reasonable for civilian consumers. For a time at least, the farm problem was solved.[24]

When the war ended, however, it became clear to agricultural policymakers in Congress, the USDA, and farmers' organizations that surpluses were again going to be a problem. In 1947 and 1949, the National Planning Association gathered together a group of agricultural economists, farm organization leaders, and labor and consumers' representatives to discuss the politics of food in postwar America. The results of these meetings, published under the titles *Dare Farmers Risk Abundance?* and *Must We Have Food Surpluses?* came to the conclusion that farmers would only continue to keep growing more and more food no matter what Congress tried to do to limit production. The only way to keep farmers from overproducing themselves into poverty, the reports argued, was to allow farm prices to rise. The key to doing this without driving up the cost of living for American consumers was, as the latter report put it, "increased efficiency in marketing to … cut costs of distribution."[25] Whereas the USDA had always focused on rationalizing the *production* of food, now they should also, according to the National Planning Association agricultural experts, use technology to rationalize the *consumption* of food.

These conferences, held by the National Planning Association, served mainly to bolster the new agricultural technopolitics already being put in place by Congress and the USDA in the mid-1940s. Of particular importance was the Agricultural Research and Marketing Act of 1946 (RMA), which explicitly ordered the USDA's economists and engineers to come up with new technologies for increasing the consumption of America's agricultural products. The bill's main sponsor, Representative Clifford R. Hope of Kansas, described the central idea of the legislation to Congress in July of 1946: "The [Research and Marketing Act] is based upon the idea of abundant production and efficient distribution and utilization of food and other farm products."[26] Efficient food distribution, according to Hope, required technologies that lowered or eliminated the cost of labor, along with technical research into the economics of mass consumption. With the optimism suggested by his surname, Congressman Hope firmly believed that more machines and smarter marketing experts would solve the farm problem, which decades worth of political haggling over how to limit production had never solved.[27] The RMA offered to create a true consensus on agricultural policy, transcending partisan divisions and uniting the interests of food producers and consumers.[28]

Three main factors led to the development of this new direction in agricultural policy. First, Congress had been taken over by Republicans in 1946 for the first time since the beginning of the Great Depression. Eager to erase the so-called socialist aspects of New Deal legislation, Republican politicians from farm states (including Hope) sought to solve the farm problem without the use of the centralized economic planning that lay at the heart of the AAA. As postwar tensions with the Soviet Union increased, price supports and acreage controls were increasingly painted as socialistic by opponents of the New Deal. This became especially clear in 1948 and 1949, when Republican congressmen and western beef ranchers spectacularly shot down the efforts of Truman's secretary of agriculture, Charles Brannan, to replace commodity price supports with direct payments to farmers based on their yearly income. Brannan had attempted to make the New Deal's agricultural policies more fair to both small farmers and consumers, but to his opponents, his attempts seemed blatantly "communistic."[29] Second, agricultural economists, like other economists, had become increasingly enamored of Keynesian theories that pointed toward steady consumption as the key to a healthy economy. The new economics stressed mass consumption rather than mass production as the key to steady growth and widespread abundance.[30] Third, there was the problem of inflation. Public opinion polls in the late 1940s and 1950s consistently ranked the rising cost of living as one of the most pressing domestic concerns.[31] Food, in particular, was constantly rising in price, and the New Deal system of guaranteeing farmers a parity price for their commodities seemed to many to be the cause. A 1951 editorial in the *New York Times*, for instance, attacked the agricultural price support system as a drag on the entire economy: "Food is the No. 1 item in the wage-earner's budget. If the price keeps rising, how can wages and the rest of the economy be stabilized?"[32] Given these pressures, even Democratic politicians from farm states became increasingly uncomfortable with the New Deal program of raising farmers' incomes by using the heavy hand of the state to raise the price of food for consumers. The new agricultural program, congressmen decided, should rely less on politically controversial economic regulations and price supports, and more on technologies of distribution and marketing.

It was one thing for the USDA's economists and engineers to receive a sharp rebuke from Congress for their previous neglect of the consumption side of the agricultural economy. It was entirely another that they received significant funding to start research projects — more than $30 million in the first five years after the RMA's passage. Agricultural engineers and economists eagerly embarked on literally thousands of research projects, studying everything from turning corn into automobile fuel to developing dehydrated food to studying the economics of air transport of grain. The majority of studies, however, focused on down-to-earth questions of how to make it cheaper for farmers and food processors to get their products to market.[33] For instance, a 1953 economic study funded by the RMA found that the cost of loading and unloading apples in warehouses could be reduced by up to 80 percent by the use of forklifts in place of belt conveyors.[34] Other RMA-funded economists sought similar technological methods for reducing costs in the marketing of milk (milk should be hauled in bulk tanks, not cans); perishable fruits and vegetables (retailers should demand careful handling in packing houses to reduce spoilage); and livestock (beef packers should modernize their stockyards to maximize the rate of feeding and slaughter).[35] Engineers, meanwhile, focused on such activities as improving corn and soybean drying and storage, using sorting machines to increase the efficiency of tomato processing plants, and developing standardized containers and packages for retail delivery of food products.[36]

Even after Congress officially cancelled the Research and Marketing Act in 1955 due to unclear results, such studies continued well into the 1960s.

This was largely because Dwight Eisenhower's secretary of agriculture, Ezra Taft Benson, created a permanently funded agency within the USDA to work on the problem of efficiently marketing food. Benson claimed to be "above politics" due to his deep Mormon faith and his training as an economist, but his actions as head of the USDA were quite explicitly aimed at erasing any vestiges of New Deal agricultural policy.[37] Immediately after taking office in 1953, he eliminated the Bureau of Agricultural Economics (BAE), claiming that too much of the BAE's economic research had focused on maintaining controversial New Deal policies such as price supports and acreage allotments (that is, paying farmers to keep some lands out of production).[38] In the BAE's stead, Benson erected two agencies, the Agricultural Research Service and the Agricultural Marketing Service. Through these agencies, he hoped to redirect the work of agricultural engineers and economists toward what he considered more "objective" marketing research.[39] Consequently, the cancellation of the Research and Marketing Act in 1955 did not end its rationale of solving the farm problem through technological efforts to streamline food marketing. In fact, projects similar to those funded by the Research and Marketing Act only became more numerous under Benson's secretaryship. For example, RMA-funded work on bulk milk hauling, begun in the early 1950s, expanded significantly when it was transferred to the Agricultural Marketing Service in the years after the RMA's cancellation.[40] Other large-scale projects of the late 1950s and 1960s studied the efficient marketing of frozen orange juice, the issue of low farm prices as a general problem in economic theory, the costs of depending on skilled laborers in food processing industries, and the proper design of food storage warehouses.[41] In hundreds of other studies, USDA economists and engineers sought to improve efficiency in marketing and distribution, covering every agricultural commodity produced in the United States, often focusing on specific regions, and often doing so in direct cooperation with food processing and retailing firms.

In the twenty years following the end of World War II, the farm problem was thus redefined. Formerly, agricultural experts, including BAE economists, farm bloc Congressmen, and Henry A. Wallace, former Secretary of Agriculture, had seen the essential problem as one of overproduction. Now, in a political culture focused on maintaining abundance without the use of "socialist" methods, the problem seemed to be one of inefficient marketing. Consequently, agricultural researchers sought ways to get more of the food consumer's dollar into the pockets of farmers. The best way to do this, argued a key set of agricultural experts, was to decrease the cost of food distribution; as we have seen above, much of this research focused on improved packaging techniques, warehousing and retailing methods, and reducing the need for skilled labor in the food marketing chain. Above all else, however, USDA economists and engineers saw the cost of transportation as the single most important area for research.

Although the cost of labor contributed the greatest increase in the price of food between the farmer and the consumer, transportation had always followed close behind labor in percentage of costs incurred in the marketing of agricultural goods and food products.[42] Unlike workers, however, transportation technologies could be reengineered with less need for political delicacy. If the cost of transportation could only be kept down, argued a 1956 USDA pamphlet meant for wide readership entitled "Food Transportation and What It Costs Us," farmers' incomes would automatically rise even as consumer prices dropped.[43]

This is where trucking came in. Ever since the late nineteenth century, farmers and their representatives in Congress had blamed the railroads for overcharging farmers to ship agricultural products. Any farmer who wished to grow wheat on the Plains, fruit or vegetables in California, or cattle in Texas was forced to depend on the railroads to get his commodities to distant urban markets. This dependence on large-scale technological systems seemed directly at odds with the agrarian ideology of farmers as independent republican producers. When the building of major highways began to make long-haul trucking a viable option for shippers in the 1930s and 1940s, agricultural leaders publicly expressed hope that truck competition would bring freight rates down. Vice President (and former Secretary of Agriculture) Henry A. Wallace, for example, addressed a group of Texas farmers and businessmen in 1943, telling them that one of Congress's main goals after the war should be to allow trucking the "opportunity to develop without suppression" to ensure "abundant consumption" of farm products. Truckers could break the monopoly of railroads on long-distance freight hauling, thereby driving down the price of getting farm products to market. Lower hauling prices would decrease the price of food for consumers, ensuring "abundant consumption" and, hopefully, preventing another agricultural depression.[44] Wallace was not an active agricultural policy maker in the postwar period, but his proposals for helping long-haul trucking become a viable competitor to the railroads would come to define a significant portion of the agricultural research undertaken by the USDA in the 1940s, 1950s, and 1960s.

The Technopolitics of Flexible Transportation

One key word — *flexibility* — summed up all that agricultural experts in the USDA imagined trucking would bring to the domestic marketing of crops, livestock, and food. Agricultural engineers, economists, and policy makers all regarded trucking as a more flexible system of transportation than railroads, but each of these groups had a slightly different idea of what flexibility entailed. For engineers, trucks could provide faster and more reliable hauling than railroads mainly because truckers were better able to provide customized hauling services. Trains hauled an incredibly diverse range of products, using a variety of railcars and switching mechanisms to move goods over long distances; but even with specialized railcars, each load was just one unit among many with widely varying needs and destinations. Each semitrailer, on the other hand, hauled only one commodity, directly from the point of origin to its destination. The commodity itself, rather than the transporter's need to limit investments in equipment, determined which type of hauling equipment would be used. Mechanically refrigerated trailers, custom livestock hauling trailers, bulk tankers for milk and oils, and grain hoppers could be designed and implemented for each specific commodity. Furthermore, truckers could provide the specialized service needed to make sure that each load arrived quickly at its destination with little damage. Truck trailers could be designed, for example, to accept standardized bulk packages of potatoes that would keep handling to a minimum during exchanges among farmers, potato processing facilities, warehouses, and retail stores. In summer, potatoes could travel in ventilated trailers to prevent degradation, while in winter they could be protected against freezing in enclosed trailers. Other commodities, from grains to livestock to dairy products to fresh fruits, had similarly customized transportation requirements that, from an engineering standpoint, trucks often seemed most capable of providing.[45]

Economists, meanwhile, tended to define trucking's flexibility in terms of systemic marketing efficiency. Railroads, in order to operate profitably, needed months of advance notice from shippers in order to allocate the appropriate number of cars to pick up a specific load at a particular time. The fickleness of climate and weather, however, has always created fluctuations in agricultural production. A farmer could only make an educated guess as to how big his crop would be come harvest time; thus, if a grain farmer ordered three railcars to arrive in the second week of October to take his grain to market, he might only be able to fill one of those cars, or might have a bumper crop that required several more cars than the railroad could provide on short notice. Truckers, on the other hand, could arrive to collect a shipment of any size with only a few days or even hours notice; from an economist's viewpoint, this "just-in-time" transportation was a much more efficient allocation of resources. As one agricultural economist summed up the issue in 1969, "Nature determines to a very large degree how much transportation will be needed, when it will be needed, and where it will be needed. ... In many cases, a saving of hours — not days or weeks — in transportation time can mean better prices for the producer or distributor, longer shelf life for the product, and better satisfied consumers."[46] Furthermore, because truckers hauled relatively smaller loads of products at greater speed than railroads, they provided food processors and supermarkets with the means to increase the rate of turnover of their products during periods of high demand. This form of flexibility was important to economists because a high rate of turnover was one of the most effective ways to assure secure profit levels for farmers and food distributors without the need to raise food prices for consumers.[47] Thus, agricultural economists tended to define trucking's flexibility in terms of efficiency. Quick, on-demand movement of a variety of goods from producers to consumers would bring stability to an otherwise constantly fluctuating food economy.

Agricultural policymakers, meanwhile, tended to conceive of trucking's flexibility in terms of competition. For policymakers of the postwar period hoping to reduce the cost of transporting agricultural products, the very presence of trucks as significant competitors to railroads promised lower freight rates for all shipments, whether by road or rail. Policy debates thus centered on how best to encourage trucking's growth, without creating a new monopolistic transportation industry by pushing railroads into bankruptcy.[48] Federally funded highway building in the 1940s and 1950s proved to be the single most important policy decision that gave truckers incentives to compete with railroads in the postwar period. Agricultural policymakers, however, had little direct influence on the development of national highway policy after the Bureau of Public Roads moved from the Department of Agriculture to the Department of Commerce in 1949.[49] The Department of Agriculture did have significant influence on other facets of transportation policy, however, particularly when it came to minimizing government regulations on truckers' geographic reach and ability to compete with other carriers. As explored in detail in the case study below, one of the most important policies in the postwar period was the agricultural exemption that allowed truckers hauling agricultural products to avoid all federal regulations but those related to safety. Agricultural policymakers also sought flexibility for truckers through reduced barriers to interstate movement. The existence of non-uniform state taxes or limitations on vehicle length and weight restricted one of the semitrailer's primary advantages over trains — their ability to move anywhere a road existed.[50] Finally, many agricultural policy analysts saw the labor situation in trucking as more flexible than that of railroading, and sought to keep it that way. Whereas railroad employees had established strong unions in the early twentieth century, most trucking firms, particularly those hauling agricultural goods, were small and non-unionized. Although relatively few truckers were self-employed, those who worked for small firms could hold out a reasonable hope of being owner-operators at some point. As a consequence, unionization was limited primarily to just a few sectors of the trucking industry, such as local and regional delivery services. To encourage this form of flexibility, agricultural economists encouraged various mechanisms that allowed small non-unionized trucking firms to compete successfully with larger firms; one key mechanism was the use of truck brokers (individuals who coordinated contracts between shippers and truckers for a commission).[51] From a policy standpoint, then, trucks were more flexible than trains because they could more easily avoid cumbersome interference from government regulators and organized labor.

The following case study of frozen food transportation in the postwar period demonstrates how these varying concepts of trucking's flexibility played out in actual practice. Interestingly, although the official justification for USDA trucking research and policy activity was to lower the cost of food transportation, the case of frozen food trucking shows that this was not necessarily the ultimate result. Because all USDA research and policy work was implemented in close cooperation with industry leaders (particularly food processors), the flexibility of trucking was often applied to purposes other than reducing transportation costs. In fact, freight rates for shipments by truck (especially over long distances) have historically been significantly higher than those for railroads.[52] However, trucking provided services that made the mass marketing of "fresh" but processed foods like frozen fruits and vegetables more reliably profitable for processors and supermarkets. USDA marketing experts believed that, with reliable profits on large volumes of goods, these food distribution firms could lower their profit margins, ultimately passing the savings backward to farmers and forward to consumers. Trucking, though itself higher priced than rail transportation, was a means for USDA researchers and policymakers to achieve a more efficient food marketing machine.

Frozen foods were the dream product for postwar agricultural experts who wanted to create a rationalized marketing machine. Once frozen, food could be held in storage in freezer warehouses for up to a year before being delivered to supermarkets. As economists saw it, the irregularity and unpredictability of nature's cycles could thus be surmounted by harvesting foods in season, then distributing them evenly throughout the year. In an ideal world, the freezing of foods would turn vegetable and animal matter into pure abstractions capable of being transported, sold, and consumed at any time or place. For example, one agricultural economist, in a fanciful aside in a technical article in 1951, dreamed of a tunnel distribution system whereby homeowners would hook up their freezers to a pneumatic tube that would deliver packages from the frozen food factory on demand.[53] Even in less fantastic visions, economists generally believed that frozen foods would significantly smooth the process of marketing foods in the postwar world. Farmers would no longer be caught with troublesome seasonal or perennial surpluses, and supermarkets would have a steady supply of quality goods to sell throughout the year. Steady incomes and profits would be the ultimate result.[54]

There was a significant technical hitch in this dream scenario, however. Most railroad refrigerator cars had been designed to keep meat and produce *cool*, not *frozen*. From the late nineteenth century until 1950, almost all railroad cars relied on either dry or wet ice to keep temperatures down. This worked fine for keeping meat chilled on the way from, say, Chicago to Boston, but it simply could not keep frozen foods *frozen*. In 1950 a cooperative effort between Frigidaire, General Motors, and the Fruit Growers Express Company brought a new kind of refrigerated railroad car on the market. Using compressed Freon, this mechanically refrigerated car could bring temperatures down to the 0 degrees Fahrenheit that frozen foods required on their trip from the farm to the supermarket to maintain color, taste, and nutrients. However, most railroads proved reluctant to invest in these expensive cars. Even when they did have these mechanical reefers, furthermore, freight trains on long hauls would make frequent stops to switch cars at rail sidings. At every stop another opportunity would present itself for frozen foods to begin melting. The frozen food industry generally regarded inadequate transportation as their single most pressing problem in the early 1950s. Frozen food packers believed that trucks, if they had reliable refrigerated trailers, could solve the problem because they traveled directly from processor to supermarket, not transferring their load until they reached the final destination.[55]

A number of private companies, particularly Fruehauf Truck Trailers, began selling reliable refrigerated trailers for hauling frozen foods in the early 1950s. Using mechanical compression refrigerant systems similar to those developed by Frigidaire, the best of these new truck trailers were able to keep temperatures well below 0 degrees Fahrenheit for sustained periods of time.[56] This was a significant improvement over the old-fashioned railroad ice cars. As a consequence, by 1957 truckers hauled nearly 80 percent of frozen food shipments.[57] But even if the new truck trailers were usually better than refrigerated railroad cars, frozen food packers continued to gripe through the 1950s that they had no way of assuring that truck drivers knew how to handle the shipments and operate the equipment properly so that foods would actually stay frozen.[58] This is where the USDA engineers, intent upon creating a flexible marketing machine for the frozen food industry, stepped in.

In 1948 a group of engineers at agricultural research stations in Pasadena and Albany, California, began a two-decade-long project dedicated to improving the distribution of frozen foods. A major part of these studies involved testing and rating refrigerated truck trailers. The engineers put a variety of commercially available trailers through rigorous testing to determine whether they would actually keep foods frozen for a reasonably lengthy trip. For example, one group of engineers loaded trailers with frozen turkeys and fruit juices, then recorded the trailer's and the food's start and end temperatures after highway trips through varying weather conditions.[59] Ten years later the engineers began publishing their results in the trade journal *Food Technology* and presenting their findings at frozen food trade conferences. The agricultural engineers had come up with a mountainous pile of data and recommendations, covering everything from the minimum thickness of insulation necessary in a refrigerated trailer (six inches) to the most efficient ways to stack pallets of frozen cargo and the precise amount of air circulation space needed to keep the goods frozen.[60] Furthermore, the engineers developed a series of rules that trucking companies could use to train their employees in the proper handling of frozen foods. Although the guidelines proved relatively expensive to implement, frozen food industry leaders saw them as absolutely necessary for maintaining consumers' trust (and therefore, profits). Frozen foods could offer remarkable quality and convenience at a relatively low price, but frozen food processors had experienced many difficult years of inconsistent quality; as a consequence, many consumers needed to be convinced to buy the products in the 1950s.[61] The recommendations developed by this task force were thus gobbled up by the frozen food industry, which credited the research with helping to make frozen foods into one of the most profitable segments of the food industry in the late 1950s and 1960s.[62] The USDA's engineers had a political mandate, a cooperative relationship with food processors, and the funding to improve the technologies of food distribution. Their work on the effective transportation of frozen foods was just one part of their effort to increase the efficiency of industrial agriculture's growing marketing machine.

Another group of USDA technical experts devoted considerable effort to keeping trucking economically competitive with the railroad industry. Of particular importance to the USDA's policymakers and marketing economists was the agricultural exemption clause of the 1935 Motor Carrier Act. This clause stated that truckers who hauled "unprocessed agricultural products" could do so without any regulation from the Interstate Commerce Commission (ICC).[63] Congress, under pressure from farm lobbyists like the Farm Bureau, originally intended the clause to apply only to individual farmers or farmer cooperatives hauling their own goods to market in small trucks. As the trucking industry grew larger in the interwar period, however, the distinction between farmers' trucks and trucks hauling farmers' products became less clear. With additional lobbying pressure from the Farm Bureau and its allies, the exemption soon covered, de facto, all trucks hauling "unprocessed agricultural products." All other truckers, meanwhile, were treated as common carriers, liable to ICC regulations on market entry and rate setting. In other words, if a trucker or trucking company wanted to haul an industrial product like automobile parts or cases of beer, he had to petition the ICC for a "certificate of convenience and necessity," which would prevent the trucker from charging unfairly low rates or encroaching on the routes of other transportation firms. However, if the trucker wanted to haul grain or hogs or milk from an Iowa farm to a processing plant, he could set his own rates, as low as he wished, and determine his own routes, without ever filing a single slip of paper with the ICC.

The USDA's marketing economists and policymakers were convinced that exempt trucking provided the flexibility needed for the efficient distribution of America's agricultural abundance. If agricultural haulers were regulated like other truckers or railroads, argued USDA marketing experts, they would be less able to adjust their routes and rates to seasonal peaks and dips in production. Larger firms with more sophisticated contracting services would then push smaller non-unionized truckers out of business. Entire regions of the country might be left with insufficient transportation services to haul the ever-increasing yields of industrialized farms. Processors who had located their factories in rural areas would be left with only the slowest and most expensive methods for hauling their products to supermarkets.[64] Consequently, from the exemption's inception in 1935 until it became irrelevant in the late 1970s when all trucking was deregulated, various Secretaries of Agriculture repeatedly fought off efforts by the ICC, regulated truckers, and railroads to eliminate the exemption.[65] Occasionally, the USDA even succeeded in liberalizing the terms of exemption. One of the most interesting of these cases involved the question of whether frozen foods could qualify as "unprocessed agricultural products."

On March 7, 1956, the United States Supreme Court sat to decide whether a chicken that had been frozen in a processing plant was still a chicken. On one side of this bizarre legal-ontological debate stood lawyers for the Department of Agriculture, contending that a frozen chicken was simply a chicken — an agricultural commodity. On the other side, lawyers for the ICC argued that a frozen chicken was a manufactured product. The USDA acted as intervening defendant on the side of Frozen Food Express, a trucking company specializing in hauling frozen foods; the ICC acted as intervening plaintiff on the side of East Texas Motor Freight Lines (a regulated carrier) and the Akron, Canton & Youngstown Railroad. The case came to the Supreme Court after the ICC ordered Frozen Food Express to cease and desist from hauling frozen poultry or meats without an ICC certificate of convenience and necessity. The trucking firm successfully appealed the ICC decision before a Federal District Court in Houston in 1951, consequently gaining the right to haul frozen foods to and from all points in the United States without restriction from the ICC. The ICC, along with several regulated carriers, appealed to the Supreme Court in 1956 out of fear that the district court's decision would set a precedent for all frozen food carriers to fall under the exemption clause.[66]

Unfortunately for the ICC, the Supreme Court sided with the secretary of agriculture on April 23, 1956, upholding the district court's decision. The Court's decision that frozen poultry should remain exempt under the Motor Carrier Act was based on the idea that a chicken's "substantial identity" was not changed by the process of freezing. This decision might seem to defy logic. A chicken that has been killed and dressed on an assembly line staffed by wage workers, then sent through a fifty-thousand-gallon vat of refrigerant solution, would seem to be about as processed as any food can be processed. However, drawing on testimony by agricultural marketing economists before Congressional committees, the Court found that the freezing of poultry was only a form of "incidental processing" necessary for making the raw chicken into a marketable product. The chicken was not manufactured by the processing plant; it was simply made into a saleable food, like milk that had been pasteurized. Furthermore, the

USDA's economists convinced the Court that the exemption was necessary "to preserve for the farmers the advantage of low-cost motor transportation."[67] As this was the expressed intent of the exemption clause of the Motor Carrier Act, the Court decided that keeping food costs low was a relevant factor in allowing frozen food carriers to operate free from ICC oversight. The *East Texas Motor* v. *Frozen Food Express* case thus set a precedent making nearly all frozen foods exempt under ICC rulings.

Frozen food packers at first celebrated the Court's decision, expecting freight rates to drop precipitously as thousands of unregulated truckers slashed their rates to corner the market on frozen foods. For instance, Seabrook Farms (located in New Jersey), one of the nation's largest frozen food packers, switched to exempt truckers immediately after the Supreme Court decision and saw their shipping costs drop accordingly.[68] Unregulated carriers tended to be small, usually owning from one to five trucks, and were able to keep their rates low because they were almost wholly unaffected by Teamsters Union wage and benefits demands. Unregulated truckers also tended to haul only full truckload shipments from one shipper to one receiver, then pick up another full truckload for the backhaul to their home base. These unregulated truckers thus avoided paying for the costly warehouse networks that allowed the larger regulated trucking companies, such as Pacific Intermountain Express or Yellow Freight, to assemble valuable less-than-truckload freight lots from multiple shippers into single delivery units aimed at one receiver.[69] In short, the Supreme Court's ruling seemed to promise frozen food packers a cheap, non-unionized freight service capable of quickly hauling their products on a moment's notice directly from factories to supermarket consumers.

In an ironic twist, however, the frozen foods industry soon found that exempt trucking was not the great blessing that economic theory had predicted. Within a year of the Supreme Court's decision, making frozen foods into "unprocessed agricultural products," representatives of the frozen food industry petitioned the USDA and Congress to draft legislation making frozen fruits and vegetables back into processed foods. This reversal of attitudes came as frozen food packers realized that they had a different definition of a flexible transportation system than the USDA's marketing economists. USDA marketing economists had defined trucking's flexibility in terms of unregulated shipping rates and unfettered geographic mobility. Frozen food packers, after experiencing this kind of flexibility in 1957, found that unregulated truckers could not be trusted to provide the quality of service necessary to keep products frozen all the way to the supermarket.[70] Most frozen-food industry leaders agreed with the agricultural *engineers'* vision of flexible transportation; trucking's primary advantage for hauling frozen foods was its specialized service, not lower freight rates. Consequently, at a 1958 Congressional hearing, frozen food representatives from Welch Grape Juice Company and Stokely-Van Camp called on Congress to return frozen foods to non-exempt status.[71]

Congress quickly complied, passing the Transportation Act of 1958, excluding frozen foods from exempt hauling status.[72]

Frozen foods were thus exempt for only a short period of time, but nearly all other agricultural products remained exempt from 1935 until all trucking was deregulated in the 1970s. Through this period, the USDA successfully maintained the exemption for commodities such as milk, grain, livestock, meat, and raw fruits and vegetables.[73] As a consequence, many small truckers were able to successfully compete with railroads and larger trucking companies in hauling America's farm products. The USDA's work in Congressional and Supreme Court debates, along with its scientific and engineering research, helped make trucks the primary method for hauling farm products to market by the 1970s. Truckers, both regulated and unregulated, hauled in 1972 at least 81 percent of meat and dairy products, 73 percent of fresh fruit and vegetables, and 99 percent of cattle and hogs.[74] Even when truckers charged higher rates than railroads per ton-mile, their speed, geographic mobility, custom service, and lack of unionization allowed farmers, food processors, and supermarkets to achieve economies of scale and scope in their marketing activities. By the 1970s the USDA's dream of a rationalized marketing machine had largely taken shape; enormous volumes of fresh foods like milk, meat, and produce traveled from farms to high-tech processing factories to supermarkets with incredible speed and remarkable uniformity in quality.

Conclusion

As a technological fix to the post-World War II farm problem, trucking probably offered little direct help in maintaining farmers' incomes without contributing to rising food prices. Although farmers' incomes did stabilize in the period from the late 1940s to the early 1970s, the main reason seems to have been that farmers who were losing money simply went out of business.[75] Surpluses continued to plague those farmers who remained, even after the 1954 passage of Public Law 480 encouraged government-subsidized sales of American agricultural products to friendly developing countries. International markets in the postwar period proved problematic as a means of eliminating surpluses, as demonstrated in the 1970s when grain sales to the Soviet Union boomed and busted in response to changing diplomatic relations, helping to cause the "Farm Crisis" of the early 1980s.[76] It is unclear, meanwhile, whether consumer food prices in the United States lowered significantly during the postwar period. The percentage of per-capita personal consumer expenditures spent on food dropped from 28 percent in 1950 to 15 percent in 1998.[77] However, this drop was mostly due to rising disposable incomes making food a smaller portion of the overall household budget; in fact, the real per-capita expenditure on food (adjusted for inflation) in the United States more than doubled between 1950 and 1998.[78] Although the new food marketing machinery of the postwar period allowed food processors and supermarkets to achieve greater economies of scale, it also allowed them to provide consumers with relatively more expensive convenience foods like frozen precooked dinners, long-shelf-life dairy products, prewashed and packaged produce, and powdered cake mixes and desserts.[79] Thus, the farm problem — maintaining farmers' incomes without increasing consumer food prices — has never really been solved. But even if trucking was not an effective solution to the farm problem itself, it was a successful technological fix in another sense.

Through its promotion of the trucking industry in the post-World War II era, the USDA successfully redefined the political-economic problem of dealing with agricultural abundance. During the New Deal, agricultural policy attempts to solve the problem of overproduction had revolved around highly publicized, hotly debated questions of political economy. Agricultural experts, along with farm organizations and other interest groups, looked to the New Deal administration to regulate the agricultural economy by setting production quotas and guaranteeing commodity prices. With the advent of the Cold War, Keynesian economics, and consumers' fears of inflation, the New Deal form of agricultural policy became unpalatable for both conservatives and liberals in Congress and in the USDA. Given these circumstances, it would seem obvious that agricultural policymakers of the 1940s and 1950s simply wanted to find less controversial ways to fix the farm problem. Focusing on technical details like the proper thickness of insulation in refrigerated truck trailers, or whether frozen foods were or were not processed, was a much safer way to work on the decades-old farm problem.

I would propose, however, that this kind of technopolitical work was not just an easy way out for politicians and bureaucrats eager to avoid controversy. Postwar American political culture was not so much conservative as it was centered on a widely shared ideology of economic abundance, fortified by Keynesian economic theory and the war experience. In a world of extraordinary abundance, the most important political questions became technical questions — namely, how to cheaply deliver the masses of industrial and agricultural products to the people who wanted them after years of privation. The USDA's work in helping 18-wheelers roll on down the highway, loaded with perfectly frozen chicken pot pies, was just one aspect of this new form of politics. Of course, like any technological fix, this solution brought a host of new problems in its wake. Truckers would come under fire in the 1970s for polluting the air, causing deadly traffic accidents, and otherwise "raising hell on the highways."[80] Furthermore, the availability of interstate highway transportation made it possible in the 1950s for food processors to move their factories out of urban centers, into the countryside where labor was cheap and non-unionized. Especially in the meatpacking industry, this ruralization of the food processing industry created a whole host of social and environmental problems that are only beginning to gain scholarly and journalistic attention.[81] The USDA largely succeeded in redefining the farm problem from a political to a technopolitical debate, but with the consequence of replacing it with what might effectively be called the "rural problem" of the early twenty-first century.

Notes

1. David Morris Potter, *People of Plenty: Economic Abundance and the American Character* (Chicago: University of Chicago Press, 1954), 122.
2. In terms of ton-miles (one ton of goods hauled over a distance of one mile), railroads continued to haul more freight than truckers through this period, due to the fact that rails carried very heavy commodities like coal, wood, and paper over very long distances. In 1972, rails hauled 42 percent of total ton-miles of freight, while trucks hauled 27.7 percent, and planes, barges, and pipelines carried 30.3 percent. Nonetheless, truckers had captured the general-purpose freight market by 1972, hauling the majority of valuable manufactured goods, food and agricultural products, building materials, and household goods. By 1972, for instance, trucks hauled 82 percent of the nation's clothing, 84 percent of office machinery, 82 percent of meat and dairy products, 66.1 percent of rubber and plastic products, 60.4 pecent of furniture, 73.1 percent of fabricated metal products, and 84.6 percent of industrial machinery. United States Bureau of the Census, *1972 Census of Transportation, Vol. 3: Commodity and Special Statistics* (Washington: Government Printing Office, 1976). Perhaps more importantly, because truckers hauled the most valuable manufactured products, they received the majority of revenue from freight hauling, earning 55.4 percent of the nation's total freight revenue in 1975 (up from 16.4 percent in 1945, when railroads earned 78.7 percent). American Trucking Associations, *American Trucking Trends* (Washington: American Trucking Associations, 1975), 17.
3. Stephen Goddard, *Getting There: The Epic Struggle between Road and Rail in the American Century* (Chicago: University of Chicago Press, 1994); Mark H. Rose, *Interstate: Express Highway Politics, 1939–1989*, 2d ed. (Knoxville: University of Tennessee Press, 1990); Frank Norris, *The Octopus: A Story of California* (New York: Doubleday, 1901).
4. William R. Childs, *Trucking and the Public Interest: The Emergence of Federal Regulation, 1914–1940* (Knoxville: University of Tennessee Press, 1985). Ellis Hawley has argued that large trucking companies, represented by the American Trucking Associations lobby group, actually desired regulation in order to prevent undue competition from independent owner-operators. Ellis W. Hawley, *The New Deal and the Problem of Monopoly: A Study in Economic Ambivalence* (Princeton, NJ: Princeton University Press, 1966), 226–46.
5. Ari Hoogenboom and Olive Hoogenboom, *A History of the ICC: From Panacea to Palliative* (New York: W. W. Norton, 1976).
6. Gabriel Kolko, *Railroads and Regulation, 1877–1916* (Princeton: Princeton University Press, 1965); Hawley, *Problem of Monopoly*.
7. The 1933 Agricultural Adjustment Act (AAA) was ruled unconstitutional by the Supreme Court because it relied for its funding on a tax on food processors. Unlike the National Recovery Act, however, the AAA was successfully re-legislated in 1938 as essentially the same program. See Kenneth Finegold and Theda Skocpol, "State Capacity and Economic Intervention in the Early New Deal," *Political Science Quarterly* 97 (summer 1982): 255–78.
8. Nelson Lichtenstein, *Walter Reuther: The Most Dangerous Man in Detroit* (Urbana: University of Illinois Press, 1995), Dana Frank, *Purchasing Power: Consumer Organizing, Gender, and the Seattle Labor Movement, 1919–1929* (New York: Cambridge University Press, 1994).
9. Lizabeth Cohen, *A Consumers' Republic: The Politics of Mass Consumption in Postwar America* (New York: Knopf, 2003); Meg Jacobs, *Pocketbook Politics* (Princeton, NJ: Princeton University Press, in press).

10. Alan Brinkley, *The End of Reform: New Deal Liberalism in Recession and War* (New York: Vintage Books, 1995); Alan Wolfe, *America's Impasse: The Rise and Fall of the Politics of Growth* (New York: Pantheon Books, 1981); Robert M. Collins, "The Emergence of Economic Growthsmanship in the United States: Federal Policy and Economic Knowledge in the Truman Years," in *The State and Economic Knowledge: The American and British Experience*, ed. Mary O. Furner and Barry Supple (Cambridge: Cambridge University Press, 1990), pp. 138–70.

11. Gabrielle Hecht, *The Radiance of France: Nuclear Power and National Identity after World War II* (Cambridge, MA: MIT Press, 1998), 15–17.

12. Brian Balogh, *Chain Reaction: Expert Debate and Public Participation in American Commercial Nuclear Power, 1945–1975* (New York: Cambridge University Press, 1991); Frank Uekotter, "Solving Air Pollution Problems Once and for All," this volume; Julian E. Zelizer, *Taxing America: Wilbur D. Mills, Congress, and the State, 1945–1975* (Cambridge: Cambridge University Press, 1998).

13. Lawrence Goodwyn, *The Populist Moment: A Short History of the Agrarian Revolt in America* (Oxford: Oxford University Press, 1978); Elizabeth Sanders, *Roots of Reform: Farmers, Workers, and the American State, 1877–1917* (Chicago, IL: University of Chicago Press, 1999); Richard Hofstadter, *The Age of Reform: From Bryan to F. D. R.* (New York: Vintage Books, 1955); Robert C. McMath, Jr., *American Populism: A Social History, 1877–1898* (New York: Hill and Wang, 1993).

14. Daniel T. Rodgers, *Atlantic Crossings: Social Politics in a Progressive Age* (Cambridge, MA: Harvard University Press, 1998), 318–66; Hal S. Barron, *Mixed Harvest: The Second Great Transformation in the Rural North, 1870–1930* (Chapel Hill: University of North Carolina Press, 1997), 81-151.

15. David Danbom, *Born in the Country: A History of Rural America* (Baltimore, MD: Johns Hopkins University Press, 1995), 161–67; Theodore Saloutos and John D. Hicks, *Twentieth-Century Populism: Agricultural Discontent in the Middle West, 1900–1939* (Lincoln: University of Nebraska Press, 1951); David A. Lake, "Export, Die, or Subsidize: The International Political Economy of American Agriculture, 1875–1940," *Comparative Studies in Society and History* 31 (January 1989): 91–94.

16. James Shideler, *Farm Crisis, 1919–1923* (Berkeley: University of California Press, 1957); John Mark Hansen, *Gaining Access: Congress and the Farm Lobby, 1919–1981* (Chicago: University of Chicago Press, 1991).

17. David E. Hamilton, *From New Day to New Deal: American Farm Policy from Hoover to Roosevelt, 1928–1933* (Chapel Hill: University of North Carolina Press, 1991).

18. Pete Daniel, *Breaking the Land: The Transformation of Cotton, Tobacco, and Rice Cultures since 1880* (Urbana: University of Illinois Press, 1985).

19. Richard S. Kirkendall, *Social Scientists and Farm Politics in the Age of Roosevelt* (Columbia: University of Missouri Press, 1966); John C. Culver and John Hyde, *American Dreamer: The Life and Times of Henry A. Wallace* (New York: W. W. Norton, 2000).

20. Culver and Hyde, *American Dreamer.*

21. Deborah K. Fitzgerald, *Every Farm a Factory: The Industrial Ideal in American Agriculture* (New Haven: Yale University Press, 2003).

22. Robert Paarlberg and Don Paarlberg, "Agricultural Policy in the Twentieth Century," *Agricultural History* 72 (September 2000): 136–61.

23. Robert C. Williams, *Fordson, Farmall, and Poppin' Johnny: A History of the Farm Tractor and Its Impact on America* (Urbana: University of Illinois Press, 1987); Katherine Jellison, *Entitled to Power: Farm Women and Technology, 1919–1963* (Chapel Hill: University of North Carolina Press, 1993).

24. On price controls, see Meg Jacobs, "'How About Some Meat?': The Office of Price Administration, Consumption Politics, and State Building from the Bottom Up, 1941–1946," *Journal of American History* 84 (December 1997): 910–41.

25. National Planning Association, *Dare Farmers Risk Abundance?* (Washington: National Planning Association, 1947); National Planning Association, *Must We Have Food Surpluses?* (Washington: National Planning Association, 1949), quote on 18.

26. *Congressional Record*, 15 July 1946, 9031.

27. House Committee on Agriculture, *Agricultural Research, Report on H.R. 6932*, 79 Cong., 2d sess., 8 July 1946; James L. Forsythe, "Clifford Hope of Kansas: Practical Congressman and Agrarian Idealist," *Agricultural History* 51 (April 1977): 407–20

28. Agricultural Research and Marketing Act, *Statutes at Large* 60, 1082 (1946); Lewis C. Mainzer, "Science in a Political Context: The Agricultural Research and Marketing Act Program," (Ph.D. diss., University of Chicago, 1957); National Planning Association, *The Agricultural Research and Marketing Act of 1946: A Consideration of Basic Objectives and Procedures* (Washington: National Planning Association, 1948); Douglas E. Bowers, "The Research and Marketing Act of 1946 and Its Effects on Agricultural Marketing Research," *Agricultural History* 56 (January 1982): 249–63.

29. Allen J. Matusow, *Farm Policies and Politics in the Truman Years* (Cambridge, MA: Harvard University Press, 1967); Kirkendall, *Social Scientists*; Barton Bernstein, "Clash of Interests: The Postwar Battle between the Office of Price Administration and the Department of Agriculture," *Agricultural History* 41 (January 1967): 45–57; Reo Millard Christenson, *The Brannan Plan* (Ann Arbor: University of Michigan Press, 1959); Virgil W. Dean, "Why Not the Brannan Plan?" *Agricultural History* 70 (spring 1996): 268–82.

30. Brinkley, *End of Reform*; Wolfe, *America's Impasse*.

31. Meg Jacobs, "Inflation: The 'Permanent Dilemma' of the American Middle Classes," in *Social Contracts under Stress: The Middle Classes of America, Europe, and Japan at the Turn of the Century*, ed. Olivier Zunz, Leonard Schoppa, and Nobuhiro Hiwatari, 130–53; "Price Index Rises to a Record High," *New York Times*, 25 July 1952, pp. 1, 15; Charles E. Egan, "Arnall Insistent Prices Will Go Up," *New York Times*, 14 August 1952, pp. 1, 24.

32. Joseph A. Loftus, "Parity Issue at Root of the Price Problem," *New York Times*, 1 April 1951, p. 151.

33. USDA, Agricultural Research Administration, *Report of Activities under the Research and Marketing Act* (Washington: GPO, 1951–53).

34. Earl W. Carlsen, *Apple Handling Methods and Equipment in Pacific Northwest Packing and Storage Houses* (Washington: USDA, Production and Marketing Administration, 1953).

35. Joseph M. Cowden, *Farm-to-Plant Milk Hauling Practices of Dairy Cooperatives* (Washington: Farm Credit Administration, USDA, 1952); Henry T. Badger, *Retail Margins for Selected Fresh Fruits and Vegetables* (Washington: USDA, Bureau of Agricultural Economics, 1953); George E. Turner and Clayton Furman Brasington, *Livestock Auction Markets in the Southeast: Methods and Facilities* (Washington: USDA, Agricultural Marketing Service, 1956).

36. *Drying Ear Corn with Heated Air* (Washington: USDA, Agricultural Research Service, 1952); A. M. Rollefson, D. B. Agnew, and C. H. Keirstead, *Improving Soybean Marketing through Farm Storage* (Washington: USDA, Production and Marketing Administration, 1951); William A. Aronow and James E. Bryan, *Prepackaging Tomatoes* (Washington: USDA, Production and Marketing Administration, 1952); E. M. Harvey, *A Comparison of Types of Containers, Refrigeration, and Loads in the Transportation of Non-Precooled Navel Oranges in Half-Box Fiberboard Cartons* (Washington: USDA, Agricultural Marketing Service, 1956).

37. Ezra Taft Benson, *Freedom to Farm* (New York: Doubleday, 1960).

38. Since its inception in the 1920s, the BAE had long relied on cultivating a relationship to commercial agricultural interest groups to support its vision of agricultural policy and planning. During the New Deal era, the BAE served as the central planning and policy wing of the USDA. Ellis R. Hawley, "Economic Inquiry and the State in New Era America: Anti-Statist Corporatism and Positive Statism in Uneasy Coexistence," in *The State and Economic Knowledge: The American and British Experience*, ed. Mary O. Furner and Barry Supple (Cambridge: Cambridge University Press, 1990), pp. 287–324; John F. Geweke et al., *Sowing Seeds of Change: Informing Public Policy in the Economic Research Service* (Washington: National Academy Press, 1999).

39. Edward L. Schapsmeier and Frederick H. Schapsmeier, *Ezra Taft Benson and the Politics of Agriculture: The Eisenhower Years, 1953–1961* (Danville, IL: Interstate Printers and Publishers, 1975).

40. Donald B. Agnew, *How Bulk Assembly Changes Milk Marketing Costs* (Washington: USDA, Agricultural Marketing Service, 1957); Bowers, "Research and Marketing Act," 262; Schapsmeier and Schapsmeier, *Ezra Taft Benson*.

41. Roy L. Lassiter, Jr. and George L. Capel, *Economic Characteristics of the Florida Chilled Citrus Juice Industry* (Washington: USDA, Agricultural Marketing Service, 1959); Imogene Bright, *The Wage Factor in Retailing Meat in 4 Cities: A Study of Marketing of Agricultural Products* (Washington: USDA, Agricultural Marketing Service, 1957); Robert K. Bogardus, *Wholesale Fruit and Vegetable Warehouses: Guides for Layout and Design* (Washington: USDA, Agricultural Marketing Service, 1961).

42. *Farm-Retail Spreads for Food Products: Costs, Prices* (Washington: USDA, Agricultural Marketing Service, 1957); Ralph L. Dewey and James C. Nelson, "The Transportation Problem of Agriculture," in USDA, *Yearbook of Agriculture, 1940: Farmers in a Changing World* (Washington: GPO, 1940), pp. 720–39; G. L. Penrose, L. T. Fuller, and J. B. Sharkey, *Transportation Spells Markets*, Bulletin 167: Oregon Agricultural Experiment Station, 1950); D. C. Wayne, *Agricultural Transportation* (Washington: USDA, Agricultural Marketing Service, 1960).

43. USDA, *Food Transportation and What It Costs Us* (Washington: USDA, Agricultural Marketing Service, 1956).

44. Henry A. Wallace, "Transportation," speech delivered 20 October 1943, Dallas, TX, in Henry A. Wallace Papers, University of Iowa Library, Special Collections, Iowa City, IA.

45. William J. Hudson and Don C. Leavens, "The Kinds and Uses of Carriers," in USDA, *Yearbook of Agriculture* (Washington: GPO, 1954), 96–7; Forrest S. Baker Jr., "Efficiency in Transportation Packaging," *Journal of Farm Economics* 46 (December 1964): 1292–94; E. P. Atrops and W. H. Redit, *Protective Services for Shipments of Carton Loads of California Oranges and Lemons* (Washington: USDA, Agricultural Marketing Service, 1962); H. D. Johnson and P. L. Breakiron, *Protecting Perishable Foods during Transportation by Truck: Meats, Fruits, Melons, Vegetables, Poultry, Dairy Products* (Washington: USDA, Agricultural Marketing Service, 1956).

46. Ivon W. Ulrey, *The Economics of Farm Products Transportation* (Washington: USDA, Economic Research Service, 1969), 1.

47. John C. Winter, "A Century of Progress," in USDA, *Yearbook of Agriculture* (Washington: GPO, 1954), 100; Donald E. Church and Margaret R. Purcell, "From Farms to First Market," in USDA, *Yearbook of Agriculture* (Washington: GPO, 1954), 87–92; Frank Groves, Richard Vilstrup, and Albert Frankenstein, *Transportation and Wisconsin's Agriculture*, Report R-2551 (Madison: University of Wisconsin, College of Agricultural and Life Sciences, 1973); David E. Moser and Wesley R. Kriebel, *Transportation in Agriculture and Business* (Columbia: University of Missouri Extension, 1964); J. K. Samuels, "The Right Product; The Right Place," in USDA, *Yearbook of Agriculture* (Washington: GPO, 1960), 276–81.

48. Hoy A. Richards, "Coordination and Competition among Carriers," in *Transportation Problems and Policies in the Trans-Missouri West*, ed. Jack R. Davidson and Howard W. Ottoson (Lincoln: University of Nebraska Press, 1967), 287–98.

49. Federal road policy prior to the 1940s focused on building farm-to-market rural roads, under the direction of "Chief" Thomas H. MacDonald at the Bureau of Public Roads. By the mid-1940s, however, advocates of intercity superhighways pushed rural road building into a subordinate position in policy debates. Bruce E. Seely, *Building the American Highway System: Engineers as Policy Makers* (Philadelphia: Temple University Press, 1987), 137–223; Rose, *Interstate*, 55–94.

50. J. S. Hillman and J. D. Rowell, *Barriers to the Interstate Movement of Agricultural Products by Motor Vehicle in the Eleven Western States*, Bulletin 248 (Tucson: Arizona Agricultural Experiment Station, 1953); H. S. Norton, *Highway Transportation Barriers in 20 States* (Washington: USDA, Agricultural Marketing Service, 1957); Josephine Ayre, *Effects of State and Local Regulations on Interstate Movement of Agricultural Products by Highway* (Washington: USDA, Agricultural Marketing Service, 1961).

51. John H. Hunter, Jr., *The Role of Truck Brokers in the Movement of Exempt Agricultural Commodities* (Washington: USDA, Economic Research Service, 1962); Walter Miklius, *Comparison of For-Hire Motor Carriers Operating under the Agricultural Exemption with Regulated Carriers* (Washington: USDA, Agricultural Marketing Service, 1966). On the cultural-economic reasons for the lack of unionization in the trucking industry, see Lawrence J. Ouellet, *Pedal to the Metal: The Work Lives of Truckers* (Philadelphia: Temple University Press, 1994); Harold M. Levinson, et al., *Collective Bargaining and Technological Change in American Transportation* (Evanston, IL: Transportation Center at Northwestern University, 1971).

52. Dwight M. Blood, "The Impact of Transportation Rate Structure on the Movement of Agricultural Commodities," *Journal of Farm Economics* 46 (December 1964): 1297–1305; American Trucking Associations, *American Trucking Trends* (Washington: American Trucking Associations, 1952–1975).

53. H. J. Humphrey, "Temperatures for Frozen Foods," *Ice and Refrigeration* 121 (August 1951): 52–58.

54. See, e.g., the testimony of USDA economist O. V. Wells in House Committee on Agriculture, Subcommittee on Consumers Study, *Frozen Food Price Spreads*, 85th Cong., 2d sess., 23 July 1958, 3–26. See also Shane Hamilton, "Cold Capitalism: The Political Ecology of Frozen Concentrated Orange Juice," *Agricultural History* 77 (Fall 2003).

55. F. Gilbert Lamb, "Let's Stress Quality in Frozen Foods," *Frozen Food Factbook*, 1957, pp. 13, 51, 53; M. J. Copley, "Keeping Frozen Foods at Zero All the Time Best Assurance of Consumer Satisfaction," *Quick Frozen Foods* (February 1957): 177; William McGinnis Holroyd, "Influences and Challenges of the Growing Frozen Food Industry on Refrigerated Transport Equipment" (Ph.D. diss., Indiana University, 1960); Wallace B. Van Arsdel, Michael Joseph Copley, and Robert L. Olson, *Quality and Stability of Frozen Foods: Time, Temperature Tolerance and Its Significance* (New York, 1969); "Refrigerator Truck Line," *Business Week*, 4 May 1946, pp. 64–5.

56. G. D. Albert, "Truck-Trailer Refrigeration for Frozen Foods," *Quick Frozen Foods* (March 1949): 98; Roy Fruehauf, *Over the Road to Progress!: Fruehauf Truck Trailers* (New York: Newcomen Society, 1957); Holroyd, *Influences and Challenges*, 44–7.

57. "Motor Carriers Haul 77.7 percent of Frozen Food Shipments," *Quick Frozen Foods* (December 1957): 128; "Truck Transportation Soars for Frozen Food Cargoes," *Quick Frozen Foods* (January 1949): 41.

58. See, e.g., W. R. Roy, "Packers' Panel on Temperatures: Concentrates," *Quick Frozen Foods* (April 1952): 61, 163.

59. "Good Results in PMA Turkey, Concentrate Truck Tests," *Quick Frozen Foods* (February 1953): 109.

60. W. B. Van Arsdel, *Food Technology* 11, 28 (1957); "Floor Racks Found Aiding Low Temperatures in USDA Tests of Truck Trailers," *Quick Frozen Foods* (April 1957): 198; John B. Hulse, "New Reefer Trailer Standards Will Guide Truck Manufacturers, FF Shippers," *Quick Frozen Foods* (December 1957): 95–103.

61. Shane Hamilton, "The Economies and Conveniences of Modern-Day Living: The Mass Marketing of Frozen Foods, 1945–1965," *Business History Review* 77 (spring 2003): 33–60.

62. "Know-How of Certified Truckers Gives FF Superior Protection," *Quick Frozen Foods* (August 1957): 31; M. J. Copley, "Keeping Frozen Foods at Zero All the Time Best Assurance of Consumer Satisfaction," *Quick Frozen Foods* (February 1957): 177.

63. Senate Committee on Interstate Commerce, *To Amend the Interstate Commerce Act, Part I: Motor Carrier Act of 1935*, Hearing, 74th Cong., 1st sess., February 25–28, March 1, 2, 4–6, 1935; Celia Sperling, *The Agricultural Exemption in Interstate Trucking: A Legislative and Judicial History* (Washington: USDA, Agricultural Marketing Service, 1957).

64. Ivon W. Ulrey, "Problems and Issues in Transportation Policy and Implications for Agriculture," *Journal of Farm Economics* 46 (December 1964): 1284; William Bredo, Robert O. Shreve, and Charles L. Hamman, *Transportation Problems of Expanding Western Agriculture* (Washington: USDA, Agricultural Marketing Service, 1954).

65. For more general works on regulation and deregulation in trucking, see Childs, *Trucking and the Public Interest*; John Richard Felton and Dale G. Anderson, eds., *Regulation and Deregulation of the Motor Carrier Industry* (Ames: Iowa State University Press, 1989); Michael H. Belzer, *Sweatshops on Wheels: Winners and Losers in Trucking Deregulation* (New York: Oxford University Press, 2000).

66. *American East Texas Motor Freight Lines, Inc.* v. *Frozen Food Express,* 351 U.S. 49 (1956).
67. Ibid, HR1 [1].
68. "Effects of Agricultural Exemption Debated by Packers and Carriers," *Quick Frozen Foods* (February 1957): 163.
69. For an excellent introduction to the complex structure of the trucking industry before the deregulation movement of the late 1970s, see Ouellet, *Pedal to the Metal,* 21-38.
70. Forney A. Rankin, "Frozen Food Exemptions Seen Stifling Carrier Service," *Quick Frozen Foods* (February 1958): 131; Philip Hertz, "Know-How of Certified Truckers Gives FF Superior Protection," *Quick Frozen Foods* (August 1957): 31.
71. House Committee on Interstate and Foreign Commerce, Subcommittee on Transportation and Communications, *Interstate Commerce Act: Agricultural Exemptions,* Hearings, 85th Cong., 2d sess., 23–25 April 1958.
72. Celia Sperling, *The Agricultural Exemption in Interstate Trucking: Developments in 1957–58* (Washington: USDA, Agricultural Marketing Service, 1959).
73. Sperling, *The Agricultural Exemption* (1957).
74. American Trucking Associations, *American Trucking Trends* (Washington: American Trucking Associations, 1975).
75. John L. Shover, *First Majority, Last Minority: The Transforming of Rural Life in America* (De Kalb: Northern Illinois University Press, 1976).
76. Jon Lauck, *American Agriculture and the Problem of Monopoly: The Political Economy of Grain Belt Farming,* 1953–1980 (Lincoln: University of Nebraska Press, 2000), 62–83; Kathryn Marie Dudley, *Debt and Dispossession: Farm Loss in America's Heartland* (Chicago: University of Chicago Press, 2000).
77. Milton C. Hallberg, *Economic Trends in U.S. Agriculture and Food Systems since World War II* (Ames: Iowa State University Press, 2001), 160–2.
78. Hallberg, 163–5.
79. Harvey Levenstein, *Paradox of Plenty: A Social History of Eating in Modern America* (New York: Oxford University Press, 1993), 100-18.
80. Robert Sherrill, "Raising Hell on the Highways," *New York Times Magazine,* 27 November 1977, p. 222.
81. Deborah Fink, *Cutting into the Meatpacking Line: Workers and Change in the Rural Midwest* (Chapel Hill: University of North Carolina Press, 1998); Brian Kirby Page, "Agro-Industrialization and Rural Transformation: The Restructuring of Midwestern Meat Production," (Ph.D. diss., University of California-Berkeley, 1993); Elizabeth Becker, "Feedlot Perils Outpace Regulation, Sierra Club Says," *New York Times,* 13 August 2002, p. 10; John W. Fountain, "Needy Workers Wait for a Kansas Plant to Reopen," *New York Times,* 10 July 2002, p. 10.

6

Synthetic Arcadias: Dreams of Meal Pills, Air Food, and Algae Burgers

WARREN BELASCO

Food is important. There is in fact nothing more basic. Food is the first of the essentials of life, our biggest industry, our greatest export, and our most frequently indulged pleasure. Food is also the object of considerable dread and anxiety. What we eat and how much (or little) we eat it may be the single most important cause of disease and death. As psychologist Paul Rozin puts it, "food is fundamental, fun, frightening, and far-reaching."[1]

Probably nothing is more frightening than the prospect of running out of food. We humans have been worrying about running out of food for a long time. Reflecting humanity's deep-rooted heritage of food insecurity, there have always been prophets warning us against complacency. And given rapid population growth and environmental degradation, it seems justified to wonder whether the banquet is over. Will our grandchildren's grandchildren enjoy the dietary abundance that most of us take for granted? And how on earth will we feed a rapidly growing, urbanized population in the Third World?

In my courses on the food system and on the future, my students always want to know how I think it'll all turn out. As a historian, I am uncomfortable making predictions. What I can do, however, is illuminate and inform the discussion by tracing and analyzing its historical evolution. Given our historical amnesia, it is all too easy to forget what has already been predicted — both rightly and wrongly.

In my book project, *Meals to Come: A History of the Future of Food* (forthcoming), I consider the way the future of the food supply has been

119

conceived and represented over the past two hundred years. I start my history of worrying in the late eighteenth century. When the economist/parson Thomas Malthus (1766–1834) published his *Essay on Population* in 1798 in response to the speculations of the Marquis de Condorcet (1743–1794) and the English radical William Godwin (1756–1836), he crystallized a three-way debate about the future of the food system.[2] As demographer Joel Cohen writes in *How Many People Can the Earth Support?* (1995) — an analysis of the carrying capacity debate — there are three enduring positions on the question of how we might feed everyone adequately in the future: (1) bake a bigger pie, (2) put fewer forks on the table, or (3) teach everyone better table manners.[3] Condorcet offered the *bigger pie* argument (1) — the techno-cornucopian position: since there are no limits on human ingenuity and creativity, science and industry can always devise ways to bake bigger and better pies for everyone. Dismissing the cornucopians, Malthus took the *fewer forks* position: humanity's capacity for reproduction outruns the farmer's capacity for production or the scientist's capacity for miracles, so prudence dictates a more conservative, less expansive approach to the future, that is, birth control and resource conservation. Somewhat pessimistic about human nature, Malthus also voiced severe doubts about Godwin's romantic-utopian *better manners* position, which held that in an egalitarian society with altruistic values, people would figure out ways to share nature's bounty and overcome scarcity. Godwin's democratic optimism was inherited and elaborated by both communists and liberals who promoted a more equitable redistribution of resources as the solution to hunger.

While the Malthusian and Godwinian positions have meshed to form the basis of modern environmentalism, with its ardent respect for carrying capacity and shared sacrifice, there is little doubt that the Cornucopian view has prevailed in mainstream American culture. As Ronald Reagan proved in 1980 and 1984, those who believe that "you ain't seen nothing yet," are far more electable than those who urge us to accept limits and boundaries. There is, as hunger analyst Pete Eisinger observes, a "presumption of abundance" in America. According to political scientist Robert Costanza, technological optimism is the "default" vision of Western civilization. Significantly, at the 1999 annual meeting of the World Futurist Society (WFS), a panel of top experts on the future of food security attracted an audience of ten people, while a panel on the future of robots attracted a hundred. Although it is tempting to dismiss such "gee whiz" dreams as symptoms of arrested development, there is no question that the giddy fantasies of youth may influence the practices of sober middle age. The attendees at the WFS robot session were largely middle-aged government workers, many of them working in high-tech agencies. As Howard McCurdy demonstrates in *Space and the American Imagination*, the American space program has spent many billions of dollars attempting to translate pulp daydreams into public policy. There is, McCurdy concludes, a strong synergy between science fiction and hard science.[4]

Mindful that technological exuberance may not always be able to distinguish the plausible from the preposterous, this paper examines three radically cornucopian technological fixes for human food insecurity: proposals to (1) reduce all meals to compact pellets (meal pills), (2) to synthesize foods from basic atmospheric elements (air food), and (3) to "industrialize photosynthesis" in automated chlorella farms (algae burgers).[5] While seemingly far-fetched, all of these ideas had serious scientific, political, and popular support in their heyday — much like genetic engineering today. Indeed, a brief scan of my endnotes reveals that these ideas were aired in many of the most respectable media of their day — eminently mainstream periodicals such as *Harper's Monthly, Scientific Monthly,* and the *New York Times,* as well as reputable think tank organs such as the *Annals of the American Academy of Political and Social Science* and the *Bulletin of the Atomic Scientists.* Proponents (and audience) included members of the scientific, political, and corporate establishment — all unified in their attraction to ideas that promised to put an end, once and for all, to humanity's archetypal challenge: hunger. Moreover, much of the appeal of these ideas stemmed from their high-tech, ultramodernistic approach to food shortages, unlike the Malthusian and Socialist alternatives, both of which would require difficult and wide-sweeping changes in human values, behavior, politics, and social structure. Why did these technological fixes fall short, and what can we learn from their ultimate failure?

First I will briefly describe and explain the appeal of these ideas, each of which posited a utopian future through chemistry — in the words of the great nineteenth-century French chemist Marcelin Berthelot, a "synthetic arcadia." Then I will suggest some common patterns and fallacies. I will devote the most attention to the meal in a pill, in part because it is the oldest and most enduring fantasy, and in part because the other two — air food and algae burgers — actually seemed more realistic (yet still very utopian) in comparison.

The meal in a pill may have the longest shelf life of the three fantasies. Despite the fact that no serious nutritionist or futurist has been advocating the meal in a pill for many years, this professionally discredited notion has retained a remarkably strong hold on popular consciousness.[6] Over and over when I ask ordinary people what they think we'll be eating in the distant future, the answer is almost always, "Oh, probably pills."[7] It's hard to pin down the precise origins of the meal-in-a-pill fantasy/nightmare. While it sounds very modern, it has archetypal roots in the basic human experience with the life-sustaining embryonic capsules called seeds. Ever since the Neolithic domestication of grains and legumes enabled humans to settle down and proliferate, mythology is full of references to minute foods with enormous powers.[8] In folklore, diminutive charms, gems, and nuggets all possess magical qualities linked to fascination with *miniaturization* and *concentration*, and we still acknowledge this "tiny is beautiful" paradigm when we speak of "pearls of wisdom" and "kernels of truth."[9] By seeking to distill vital essences into a highly condensed "elixir of life," food-pill advocates also echoed ancient dreams of *replication* (the invention of manmade surrogates for nature), and *alchemical transformation* (the conversion of base substances — here, primary chemicals — into precious, life-sustaining food). (These archetypes are also seen in the dreams of manufacturing food from air and microbes.) And like air food and algae burgers, the meal in a pill offered a magical fix — an elegantly simple solution to that most timeless and universal of human problems, the struggle for food. Reducing feeding to pill popping would seem to settle once and for all those intractable issues of agriculture, hunger, health, and household labor that have plagued humanity for millennia. Combining several archetypes, a 1936 *Popular Science Monthly* article on "Life from the Test Tube," predicted that "modern alchemists" of the food laboratory would soon condense the "elixir of life" into "food pills that would contain everything necessary for life — a feat that would render man forever independent of natural resources for his nourishment, and banish fear of crop failure and famine.[10]

While this search for the ultimate insurance of food security — and complete freedom from natural limits — was as old as Prometheus, the actual meal-in-a-pill idea version seems to have surfaced in the late nineteenth century, a period of enormous economic growth, booming population, significant scientific and technological innovation, and rapid changes in everyday social life, especially among the middle-class urban people who were the audience for the popular literature where the idea was discussed most frequently, primarily in the context of reformist speculation.[11] Perhaps one reason why the meal pill fantasy had "legs" is that, along with alleviating dread shortages, it also offered positive advantages of convenience, efficiency, and health. That is, in addition to seeming the ultimate solution to food shortages, the meal pill also addressed four other concerns of the time: heightened geographical mobility, the "woman question," the progressive war on waste, and dietary reform.

Mobility: With the population boom, Europeans took to the road and seas in unprecedented numbers, and pioneering food technologists sought solutions to the logistical problems of migration and conquest. Thus, inspired by the Donner Party disaster of 1846 (when stranded California Sierra pioneers resorted to cannibalism), Gail Borden vowed to create an array of concentrated foods to save future migrants: "I mean to put a potato into a pillbox, a pumpkin into a tablespoon, the biggest sort of watermelon into a sauce. The Turks made acres of roses into attar of roses. I intend to make attar of everything."[12] Borden never got to potato pills, but he did develop concentrated meat biscuits, which proved unpalatable, and then condensed milk (1856), one of the first great triumphs of modern food processing and the precedent for other marketing break-throughs purporting to cover all nutritional bases in one convenient dish: for example, Kellogg's breakfast cereals, Liebig's infant formulas, Fleischmann's yeast cakes, and Campbell's soup, the last long advertised as "a whole meal in one soup!" Extending the concentrating tendencies suggested by canning, hardtack, and beef bouillon, the pill promised the complete culinary portability deemed essential for maximum geographical mobility. And with outer space looming as the ultimate solution to population growth, projections of future space travel also included food pills as potential rations for multiyear journeys.[13]

Feminism: Victorian feminists generally agreed on the need to reduce housework as a way to expand women's freedom. Some utopians organized cooperative housekeeping networks, fantasized about fully automated kitchens, or advocated centralized commissaries with home delivery services.[14] Others took domestic reform several steps further by dreaming of a complete divorce from cooking and dining through synthetic foods, especially the meal in a pill. For example, in 1893 feminist/populist agitator Mary E. Lease predicted that by 1993 agricultural science would allow us "to take, in condensed form from the rich loam of the earth, the life force or germs now found in the heart of the corn, in the kernel of wheat, and in the luscious juice of the fruits. A small phial of this life from the fertile bosom of Mother Earth will furnish men with substance for days. And thus the problems of cooks and cooking will be solved."[15] In *The Republic of the Future*, Anna Dodd's 1887 satirical novel about New York in 2050, pneumatic tubes delivered prescription bottles of food tablets directly to kitchenless apartments. While Dodd was a social conservative who deplored feminist aspirations, her spoof closely mimicked the sober tone of the period's reigning utopian literature, which was usually quite serious in addressing the need to "revolutionize" domestic work. Thus, with tongue in cheek, Dodd put the case succinctly: "When the last pie was made into the first pellet, women's true freedom began."[16]

Efficiency Engineering: These reformist impulses dovetailed with an emerging faith in enlightened engineering and centralized, "scientific" management. If these progressives had a common enemy, it was waste — the waste of time, energy, resources, and human potential. Conversely, their common goal was to maximize efficiency by rationalizing and consolidating every activity, including housework, shopping, and dining. In addition to developing the modern welfare state, with its central bureaus, this was the era that developed the department store, the modern university, the cafeteria, the public museum, and the amusement park — all amalgamating institutions offering a convenient, one-stop consumption experience under one roof.[17]

In a sense the meal in a pill represented a culmination of the technological utopian's embrace of comprehensive, all-in-oneness. The fact that meal pills might actually reduce food choices enhanced their efficiency value, for this was the era that produced Henry Ford, whose celebrated Model T came in "every color as long as it's black." Just as the Model T expressed the one size fits all mentality of early mass production, the meal in a pill embodied the efficiency engineer's determination to reduce costs by concentrating and streamlining essential life processes.[18]

For example, in the 1899 novel, *Looking Forward: A Dream of the USA in 1999*, Arthur Bird predicted that scientific management would combine with capitalist enterprise to produce "Ready Digested Dinners." Fulfilling the Taylorist emphasis on prudent time management, meal pills would free busy modern workers from wasteful, lengthy lunches, and the naps that came after heavy lunches. "In order to save time, people [in 1999] often dined on a pill — a small pellet which contained highly nutritious food. They had little inclination to stretch their legs under a table for an hour at a time while masticating an eight-course dinner. The busy man of 1999 took a soup pill or a concentrated meat-pill for his noon day lunch. He dispatched these while working at his desk." As was often the case with technological utopians, Bird combined far-out gadgetry with social conservatism. Thus, in line with inherently patriarchal assumptions, his "fair typewriter" of 1999 preferred feminine "ice-cream pills" and "fruit pellets" to the more masculine "bouillon or consommé pellets."[19]

Scientific Eating: There was a surprisingly short gap between pulp fiction and the futuristic visions of real scientists, particularly the "scientific eating" crusade.[20] This turn-of-the century movement reflected and reinforced technological utopianism with claims that because science and industry, not tradition and nature, would soon determine what people ate, a diet of meal pills and other synthetic foods (like air food and algae burgers) was not at all implausible.

For one thing, the reigning chemical paradigm (established by Justus von Liebig by the 1850s) insisted that all foods could be reduced to a few basic elements. So complete was the chemist's reductionist victory that, as early as 1860 Ralph Waldo Emerson — not otherwise known for modernist affections — could pronounce: "Tis a superstition to insist on a special diet. All is made at last of the same chemical atoms."[21] Since nitrogen is nitrogen, carbon is carbon, bold chemists claimed that man could synthesize anything from basic chemical elements. After developing artificial fertilizers, Liebig also worked on early infant formulas to replace breast milk. As Liebig's successor, the renowned French chemist Marcelin Berthelot put it in an 1894 preview of "Foods in the Year 2000," "when the milk has left the cow, it is merely a chemical compound and with it physiology has nothing to do." Illustrating the technological utopian's penchant for infinite extrapolation of present-day inventions, Berthelot reasoned that because synthetic butter (margarine), sugar (saccharin), vanilla (vanillin), and indigo (alizarin) were already on the market, beefsteak in tablet form (perhaps made from coal) was probably just around the corner. "We shall give you the same chemical food, chemically, digestively, and nutritively speaking. Its form will differ, because it will probably be a tablet. But" — in a concession to the emerging capitalist ideology of consumer sovereignty — "it will be a tablet of any color and shape that is desired and will, I think, entirely satisfy the epicurean senses of the future."[22]

Well-aware of mounting concerns about food adulteration and contamination, Berthelot argued that synthetic foods would be both tastier and safer than the natural variety." Strange though it may seem, the day will come when man will sit down to dine with his toothsome tablet of nitrogenous matter, his portions of savory fat, his balls of starchy compounds, his casterful of aromatic spices, and his bottles of wines or spirits which have all been manufactured in his own factories, independent of irregular seasons, unaffected by frost, and free from the microbes which over-generous nature sometimes modifies the value of her gifts." In line with mounting pressure for environmental conservation in the more industrialized nations, Berthelot also put a back-to-nature spin on his vision of a "synthetic arcadia": by eliminating agriculture, synthetic foods would return the overfarmed countryside to wilderness and save natural resources. "If the surface of the earth ceases to be divided by the geometrical devices of agriculture, it will regain its natural verdure of woods and flowers ... The favored portions of the earth will become vast gardens, in which the human race will dwell amid a peace, a luxury, and an abundance recalling the Golden Age of legendary lore." Coming from the acknowledged "foster-father of synthetic chemistry" (Liebig was the primary patriarch), Berthelot's prophecy was much quoted and debated over the next forty years, and his argument that only chemistry can "save" nature is still widely repeated by proponents of high-tech food production.[23]

Even though the meal in a pill sounds so far-fetched, even preposterous, I've devoted a lot of space to it because it set the pattern for other technological fixes — the strong archetypes; the high hopes for ultimate solutions to pressing problems and concerns; the exuberant extrapolation of current developments; the appeals to efficiency, reason, and environmentalism; the cross-influences between science fiction and scientific research. Also, on the spectrum of futuristic foods, the pill established the most extreme, far-out pole compared to which other technologies seemed *less* far-fetched or preposterous. Thus, the idea that foods might be synthesized directly from the elements in the air (air food) surfaced at the end of the nineteenth century as a seemingly more reasonable possibility than tablet foods, for these artificially derived substances would at least resemble natural foods, much the way margarine looked like butter. And compared to air food, the notion of harvesting proteins and fat from algae and then converting them to burgers, shakes, and animal feed (an idea that surfaced in the 1920s and peaked in the 1950s) seemed so downright practical that it received considerable funding from several important foundations and universities.

Just as the meal in a pill was an enthusiastic extrapolation of existing technological breakthroughs in food manufacturing (e.g., bouillon cubes, condensed soups, fortified breakfast flakes), air food extended turn-of-the-century hopes for synthesizing nitrogen fertilizers directly from the air. In 1898 the British scientist William Crookes had issued one of the direst Malthusian warnings — the prediction that, as the world's natural supply of nitrates ran out, the world's wheat farmers would not be able to meet the needs of a growing population. "England and all civilised nations stand in deadly peril of not having enough to eat," Crookes had stated, in classic wake-up call rhetoric.[24] Soon afterwards, however, popular science journals were talking up the Haber Process, which in effect used electricity to synthesize ammonia, a building block of fertilizers, from "atmospheric nitrogen."[25] Although this was, in fact, a very energy-intensive process, it seemed as if science had been able to find a free source of a vital resource — or *almost* free given that, according to the boosters of electrification, the price of electricity would soon drop to near zero with the imminent perfection of new power sources, whether hydroelectric, solar, radio waves, nuclear, or the inevitable next discovery. When two seemingly unlimited resources are paired — air and electricity — the result is the euphoric hype of infinite expectation. "With boundless atmospheric nitrogen and with water power [to provide the electricity needed to "fix" it] … man need never fear an insufficient nitrogenous food supply," one Iowa state scientist exulted in 1911.[26] If humanity could at last be liberated from the need for fertile soil, an era of universal peace surely beckoned. After all, as one 1902 *Harper's* discussion of the nitrogen research had noted, "the effort to supply this one-hundredth of the plant's food has caused most of the wars and conflicts of the world …"[27] And the fact the process was further improved as a by-product of German weapons research added to the archetypal appeal of this "swords into ploughshares" story. Haber's discovery, farm editor Wheeler McMillen wrote, "is one of the most momentous of the modern era. It sets forward for centuries the fulfillment of the Malthusian prediction."[28] To be sure, a 1924 survey of "Our Nitrogen Problem" found that, all the hype notwithstanding, atmospheric nitrogen was actually made in just "two little plants" that were "so small as to have no effects" on the market, which still relied primarily on imports of Peruvian guano and Chilean nitrates. And as economist Joseph Davis noted in 1932, global markets were soon awash in wheat *not* because of artificial fixation, which remained a potential nitrogen source, but because world wheat production had expanded the old-fashioned, imperialistic way, by plowing up new land, much of it unsuitable to long-term intensive cultivation.[29] But even amidst the ensuing Dust Bowl, the happy lesson of atmospheric nitrogen was repeated in cornucopian surveys of future food production.

Moreover, as was so often the case in hopeful speculation, one success story inspired the expectation of many more. If fertilizer could be synthesized directly from the nitrogen in air, could harvesting food directly from the elements be so far behind? Echoing Berthelot's boasts from the 1890s, some chemists claimed that they would soon be able to synthesize basic food elements from coal (an obvious source of carbon) or, again, even from the air, which had great quantities of nitrogen, a basic building block of protein. A 1907 *Everybody's* article on the "miracle of synthetic chemistry" predicted that, based on the recent successes with nitrogen fixation, scientists would soon want to make "a loaf of bread or … a beefsteak" from " a lump of coal, a glass of water, and a whiff of atmosphere."[30] Such forecasts had an aura of inevitability by the 1920s when, as Jeffrey Meikle shows in *American Plastic*, the "abstract Chemist assumed heroic proportions" — a veritable "master of 'the science of the transformation of matter.'"[31] But while almost all experts agreed that synthetic foods were plausible and perhaps inevitable, they disagreed sharply about the ultimate source. While many allowed that food could theoretically be synthesized from air — nitrogen was nitrogen after all — the more "realistic" prognosticators looked to more earthly sources, especially yeast and algae nurtured on cellulose, a major "waste product" of agriculture, forestry, and many industrial processes.

Such plans to boost food production tended to stall during the Depression of the 1930s, when hunger increased despite huge agricultural surpluses and a lower birth rate — dramatic support for the egalitarian/socialist position that hunger was the result of inequitable distribution, not overpopulation or inadequate supply. But World War II once again raised neo-Malthusian fears of population pressure and of famine not only in Europe but also in the Third World, where life-saving medicine and pesticides brought what was called the population explosion or bomb, which some saw as a greater threat to world peace than the atomic bomb. To this base the ensuing Cold War added fears that the communist second world would exploit growing hunger in the supposedly overpopulated third.[32] With socialistic analysis of distribution largely excluded and with the Malthusian push for birth control still too controversial as well, cornucopians tended to dominate mainstream discussion of future food needs with their proposals for massive industrial development, high doses of agrichemicals, peaceful applications of nuclear power, and the invention of new food sources.[33]

High on the short list of promising new technologies was algae. Measured against the ultramodernistic bravado of meal pills and air food, the new algae research of the 1940s actually seemed a more moderate, even natural approach, for it came not from the brave new world of synthetic chemistry, but from the more familiar world of marine biology. For years naturalists had been urging a closer look at unexploited sea resources as another way out of the Malthusian trap. Why bother concocting artificial meals from coal when there was so much natural seafood out there just for the taking? While discussion of marine resources focused mainly on fish farming (aquaculture) and on using unappreciated "trash fish," mollusks, and mammals, the more ambitious proposals also targeted highly nutritious kelp (macro-algae) and plankton — the floating stew of micro-algae and tiny animals that constitute the base of the food chain.[34] And then in 1948, at the height of the latest population bomb scare, alluring news came from pilot projects sponsored by the Carnegie Institution and conducted by the Stanford Research Institute in Palo Alto, California, and Arthur D. Little, Inc. in Cambridge. Initial results suggested that chlorella algae was an astounding photosynthetic superstar. When grown in optimal conditions (sunny, warm, shallow ponds fed by simple carbon dioxide), chlorella converted upward of 20 percent of solar energy (vs. the conventional wheat field's 1 percent) into a plant containing over 50 percent protein when dried. Unlike most plants, chlorella's protein seemed complete, for it had the ten amino acids then considered essential, and it was also packed with calories, fat, and vitamins.[35]

Since these results were very preliminary, researchers hedged the early scholarly reports with caveats about the need for a lot more work. Yet even the most responsible investigators were unable to resist heady extrapolation. Thus, Dean Burk, a National Institutes of Health scientist actively involved in the Stanford pilot, speculated that scientists would soon be able to quadruple chlorella's already impressive photosynthetic efficiency.[36] Based on a mere 100 pounds of chlorella produced at the Cambridge pilot plant, Carnegie's scientists projected possible yields of 17 to 40 tons (dry weight) per acre. This meant 17,000 to 40,000 pounds of protein per acre compared to 250 to 800 pounds of protein from soybeans, the most efficient conventional plant, and all at a reasonable cost of 25 cents per pound. Taking the lower yield, a plantation the size of Rhode Island would be able to supply half of the world's daily protein requirement of 65 grams.[37]

And these yields were low compared to other estimates. One scientist predicted possible yields of 55,000 pounds of protein per acre, with virtually insignificant harvesting costs.[38] Another estimated that one 1,000-acre chlorella farm, staffed by just 20 workers, could produce 10,000 tons of protein per year — all at a cost of just $10 per ton; taking one more step, 50 million tropical acres devoted to algae production would *double* the world's food supply. Quoting Dean Burk, a jubilant front page article in the *New York Times* estimated algae yields of 200 tons an acre — one hundred times the average yield of an acre of conventional crops.[39]

If scientists could be so fast and free with the figures, then journalists felt licensed to draw full scenarios. A 1948 *Collier's* article, "Food Pumped from Pipelines," sketched the "farm of the future" (just twenty-five years ahead) to be located somewhere on the shoreline of the Texas Gulf, southern California, Galilee, or some other sunny place: "For miles — as far as the eye can see — twist fat coils of glass pipe, two and three feet thick. Greenish-yellow fluid courses sluggishly through the transparent mains ...

Great pumping stations dot the shore. The coils lead inland and terminate in a huddle of other plants from which a mile-long train of tank cars — the only sign of human life — is just gliding away." In the sun-drenched tubes billions of algae — "microscopic factories" — multiplied "to produce edible foods — butter, tissue-building proteins, starches, fodder for cattle." Hardly a human hand would be involved, as "automatically operated devices, requiring only switchboard control, swirl the harvest through batteries of centrifuges" separating "crop" from water before pipelining it to "solvent extraction tanks for processing into food for man and beast." Although the scene had "the appearance of a Dali surrealist landscape," the implications were anything but absurd: "In the farm of the future … man has learned how to grow food in water in such abundance that, for the first time in history, there is enough for everybody, and hunger is banished from the earth" — a biotechnological utopia.[40]

Of course it never happened, nor did the other versions of a synthetic arcadia. Instead, we eat foods grown in soil, converting sunlight and minerals to nutrients in the same, old-fashioned, "inefficient" way. To be sure, more people are better fed (or stuffed) than ever before, but this progress has come through relatively minor tinkering with nature's timeless photosynthetic processes. While we have vastly increased the inputs, especially of energy-intensive chemicals and fuels, such improvements seem rather small adjustments when compared with the radical streamlining and simplification envisioned by the technological-utopians. So what happened? In my remaining space I would like to suggest a few general pitfalls of these technological fixes.

Economics: Beware of seemingly revolutionary efficiencies. They usually underestimate the costs, especially the energy inputs (as in air food and algae). (This is, of course, a major problem in calculating the full costs of the "cheap" food produced by industrial monoculture, which, by some calculations may turn out to be far less efficient than diversified small farms.[41]) And they also underestimate the ability of existing technologies to out-compete them. Thus it proved much cheaper to produce protein from soybeans, grown in conventional soil, than to harvest it from the air or algae ponds.

Political Economy: Beware of any innovation that purports to eliminate a major industry, especially the largest one. The food industry is too big, too entrenched, and too politically powerful to allow itself to be replaced by a feeding system that is so rationalized as to threaten millions of jobs.

Warring Archetypes: As I have noted, synthetic foods had deep roots in the mythology of alchemy and wizardry, but opponents also have their archetypes. For every noble Prometheus overcoming natural limits there is an overreaching Frankenstein inviting retribution. Historically speculative fiction has both inspired technological innovation and, through its dystopian variants, constrained it. The film *Soylent Green* may have done more to undo the algae burger than any scientific or economic obstacle.

Similarly, as we have seen recently, labels like "Frankenfood" and "Brave New Farm" have proved to be powerful hindrances to popular acceptance of biotechnology.

Reductionism: Each of the technological fixes discussed here greatly oversimplified the causes of and solutions for human suffering. First, while we can understand why food scientists might reduce all human conflict and hardship to a shortage of food, much research shows that hunger and famine are not the result of food shortages — at least not on a global level.[42] We also know that people can be mean, greedy, and violent even when they have lots of food. And what do we mean by *food* anyway? By reducing the complex array of diet-related activities and behaviors to the biochemical appropriation of a few nutrients, especially protein and calories, each of the three fixes took an exceptionally narrow view of food. With the discovery of more and more phytochemicals, we're only beginning to understand that good nutrition entails a wide variety of ingredients. And that's only the start when it comes to appreciating the importance of food. And at some point every food scholar must quote Roland Barthes to the effect that food is far more than nutrients; it is also closely tied up with our sense of who we are, where we come from, and how we relate to people.[43] The production, exchange, and sharing of food is so central to group membership that it would be socially disastrous, or at least very destabilizing, to eliminate the rituals of eating. Indeed, even NASA, which once defined eating as the ingestion of "edible biomass," has conceded that astronauts on long-term missions "like to eat at least one of their three daily meals together to socialize and build camaraderie."[44] People still like taste, they like variety, and they like amplitude and bulk. Notwithstanding the mounting demand for convenience, people like to work a bit on their food, at least to the extent of heating and chewing. Food marketers and engineers have also concluded that consumers favor products that resemble what is considered to be a traditional form. Even space food has moved in that direction — away from synthetic pouch mush to fajitas, pasta, and curries.[45] Indeed, it is increasingly clear that many people like to think that their food has an identifiably natural origin (and not a biochemist's definition of *natural*) — a grazing cow, a rooted plant, not a free-floating molecule or cell. Berthelot's extrapolative leap — that once people accepted something "artificial" like margarine, they would accept *anything* "artificial" — has not played out, yet. As the chastened biotech companies have learned, success in consumer capitalism comes through catering to a fundamental ambivalence toward modernism. Futurist bravado notwithstanding, most people want the convenience of the future, and the look, taste, and sensibility of the past.[46] The same applies to feeding the hungry masses. It is a mark of modernist hubris and arrogance to believe that the poor will be any more amenable to meal pills, air food, or algae than the rich.

Notes

1. Paul Rozin, "Food is Fundamental, Fun, Frightening, and Far-Reaching," *Social Research* 66 (winter 1998): 9–30. For an extended meditation on food's multiple meanings: Warren Belasco, "Food Matters: Perspectives on an Emerging Field," in *Food Nations: Selling Taste in Consumer Societies*, ed. Warren Belasco and Philip Scranton (New York: Routledge, 2002), 2–23.
2. Thomas Malthus, *An Essay on the Principle of Population*, ed. Andrew Flew (London: Penguin, 1985.)
3. Joel E. Cohen, *How Many People Can the Earth Support?* (New York: Norton, 1995), 17.
4. Peter K. Eisinger, *Toward an End to Hunger in America* (Washington, DC: Brookings, 1998), 3–5; Robert Costanza, "Four Visions of the Century Ahead," *The Futurist*, February 1999, 23; Howard McCurdy, *Space and the American Imagination* (Washington, DC: Smithsonian, 1997).
5. Portions of the paper are adapted from Warren Belasco, "Algae Burgers for a Hungry World? The Rise and Fall of Chlorella Cuisine," *Technology and Culture* 38 (1997): 608–634; Warren Belasco, "Future Notes: The Meal-in-a-Pill," *Food and Foodways* 8 (2000): 253–271.
6. A rare exception is the 1999 prediction by Manfred Kroger, professor of food science at Pennsylvania State University, that within 100 years, "it's possible people won't eat meals at all but instead will consume individually designed wafers with just the right amount of fat, protein, fiber, carbohydrates, vitamins, minerals and medicinal herbs to meet their individual needs." Nancy Hellmich, "Future Food Could Lengthen Your Life," USA Today, 5 January 1999, http://www.usatoday.com/hotlines/diet/diet001.htm, accessed June 28, 1999.
7. Although not based on scientific polling, this observation is drawn from several hundred open-ended interviews and discussions concerning the future of food with colleagues, students, friends, and strangers. A similar pattern prevails in a recent Delphi-style survey of opinions concerning the dinner of the future. While several dozen food scholars, writers, and professionals divided between Malthusian and Cornucopian visions of the year 2050, the bluntest and possibly most vernacular opinion was offered by an octogenarian counterman at New York's Carnegie Delicatessen: "There will be nothing on the plate. You'll eat a pill." "What's for Dinner in 2050?" *Beard House* (spring 1999), 36.
8. On the centrality of grains: Margaret Visser, *Much Depends on Dinner* (New York: Collier, 1986), 22–55, 155–191; Jared Diamond, *Guns, Germs, and Steel* (New York: Norton, 1999), 85–190. Maguelonne Toussaint-Samat, *History of Food* (Oxford: Blackwell, 1994), 45–56, 128, 247–290; Harold McGee, *On Food and Cooking* (New York: Collier, 1984), 249; Maria Leach, ed., *Dictionary of Folklore* (New York: Funk & Wagnalls, 1949), 123. More recently, the potato has had a similar role as an all-in-one miracle food. See William McNeill, "How the Potato Changed the World," *Social Research* 66 (spring 1999): 67–83.
9. It is probably not coincidental that the word "tablet" is the diminutive of "table," originally a board covered with food. Etymologically speaking, a tablet might thus be considered a mini-dinner.
10. Robert E. Martin, "Life from the Test Tube Promised by New Feats of Modern Alchemists," *Popular Science Monthly*, June 1936, 15.
11. On the primarily middle-class audience for speculative thought: I. F. Clarke, *The Pattern of Expectation, 1644–2001* (New York: Basic Books, 1979); Joseph J. Corn, ed., *Imagining Tomorrow: History, Technology, and the American Future* (Cambridge: MIT Press, 1987); Howard P. Segal, *Technological Utopianism in American Culture* (Chicago: University of Chicago Press, 1985). For an overview of the stresses and dreams of late-nineteenth-century, middle-class American culture: Harvey Green, *The Light of the Home* (New York: Pantheon, 1983); T. J. Jackson Lears, *No Place of Grace: Antimodernism and the Transformation of American Culture, 1880–1920* (New York: Pantheon, 1981); James Gilbert, *Perfect Cities: Chicago's Utopians of 1893* (Chicago: University of Chicago Press, 1991).
12. Waverley Root and Richard de Rochemont, *Eating in America*, (New York: William Morrow, 1976), 159.
13. Jack Goody, "Industrial Food," in *Food and Culture*, ed. Carole Counhihan and Penny Van Esterik (New York: Routledge, 1997), 338–356; John F. Mariani, *The Dictionary of American Food and Drink* (New York: Hearst Books, 1994), 256–257.

132 • Warren Belasco

14. Dolores Hayden, *The Grand Domestic Revolution: A History of Feminist Designs for American Homes, Neighborhoods, and Cities* (Cambridge, MA: M.I.T. Press, 1981); Polly Wynn Allen, *Building Domestic Liberty: Charlotte Perkins Gilman's Architectural Feminism* (Amherst: University of Massachusetts Press, 1988).

15. Mary E. Lease, "Improvements So Extraordinary the World Will Shudder," in *Today Then: America Best Minds Look 100 Years into the Future on the Occasion of the 1893 World's Columbian Exposition*, ed. Dave Walter (Helena, MT: American & World Geographic Publishing, 1992), 178.

16. Hayden, *Grand Domestic Revolution*, 134; Jean Pfaelzer, *The Utopian Novel in America: 1886–1896* (Pittsburgh, PA: University of Pittsburgh Press, 1984), 81–83. For an introduction to late-nineteenth-century utopian fiction: Carol Farley Kessler, ed., *Daring to Dream: Utopian Stories by U.S. Women: 1836–1919* (Boston, MA: Pandora Press, 1984); Kenneth Roemer, *The Obsolete Necessity: America in Utopian Writing, 1888–1900* (Kent, OH: Kent State University Press, 1976).

17. Edward Bellamy, *Looking Backward* (New York: New American Library, 1960 [orig. 1888]); Howard P. Segal, *Technological Utopianism in American Culture* (Chicago: University of Chicago Press, 1985); Cecilia Tichi, *Shifting Gears: Technology, Literature, and Culture in Modernist America* (Chapel Hill: University of North Carolina Press, 1987); Carolyn Marvin, *When Old Technologies Were New: Thinking About Electric Communication in the Late Nineteenth Century* (New York: Oxford, 1988); Harvey Levenstein, *Revolution at the Table* (New York: Oxford, 1988) 44–59, 98–108; Alan Trachtenberg, *The Incorporation of America: Culture & Society in the Gilded Age* (New York: Hill & Wang, 1982); Bradford Peck, *The World a Department Store: A Twentieth Century Utopia* (Lewiston, ME: Bradford Peck, 1900); Martha Banta, *Taylored Lives: Narrative Productions in the Age of Taylor, Veblen, and Ford* (Chicago: University of Chicago Press, 1993). See Marinetti's *Futurist Cookbook* (San Francisco: Bedford Arts, 1989) for a reductio ad absurdum of the efficiency argument.

18. On Fordism: James J. Flink, *The Automobile Age* (Cambridge, MA: The M.I.T. Press, 1988), pp. 40–55.

19. Arthur Bird, *Looking Forward: A Dream of the United States of the Americas in 1999* (New York: Arno, 1971 [orig. 1899]), 185.

20. Levenstein, *Revolution at the Table*, pp. 72–160. For a parallel case of the interaction between science fiction and real science, see Howard E. McCurdy, *Space and the American Imagination* (Washington, DC: Smithsonian, 1997).

21. Emerson quoted in *Since Eve Ate Apples*, ed. March Egerton (Portland, OR: Tsunami Press, 1994), 253. On the reductionist paradigm: Ross Hume Hall, *Food for Nought: The Decline of Nutrition* (New York Vintage, 1974); Elmer McCollum, *A History of Nutrition* (Boston: Houghton Mifflin, 1957), 84–99.

22. Henry J. W. Dam, "Foods in the Year 2000," *McClure's Magazine*, September 1894, 311, 306. On Liebig: Levenstein, *Revolution at the Table*, 311

23. Dam, "Foods in the Year 2000," 312. On saving wilderness through high-tech farming: Walter Truett Anderson, "Food without Farms: The Biotech Revolution in Agriculture," *Futurist*, January–February 1990, 16–21; Paul E. Waggoner, *How Much Land Can Ten Billion People Spare for Nature?* (Ames, IA: Council for Agricultural Science and Technology, 1994).

24. Joseph S. Davis, "The Specter of Dearth of Food: History's Answer to Sir William Crookes," in *Facts and Factors in Economic History*, ed. Edwin Francis Gay (Cambridge: Harvard, 1932), 733.

25. Ray Stannard Baker, "The Scientist and the Food Problem," *Harper's Monthly Magazine*, November 1903, 932–937; F. G. Cottrell, "Fertilizers from the Air," *Scientific Monthly*, September 1925, 245–249.

26. Elbert W. Rockwood, "The Work of the Chemist in Conservation," *Popular Science Monthly*, March 1911, 295.

27. Baker, "The Scientist and the Food Problem," 934.

28. Wheeler McMillen, *Too Many Farmers: The Story of What Is Ahead in Agriculture* (New York: William Morrow, 1929), 205.

29. Harry A. Curtis, "Our Nitrogen Problem," *Annals of the American Academy of Political and Social Science*, March 1924, 173; Davis, "The Specter of Dearth of Food," 741.

30. Henry Smith Williams, "The Miracle-Workers: Modern Science in the Industrial World," *Everybody's Magazine*, October 1907, 497–498.

31. Jeffrey I. Meikle, *American Plastic: A Cultural History* (New Brunswick, NJ: Rutgers University Press, 1997), 31.

32. William Vogt, *Road to Survival* (New York: William Sloane, 1948); Fairfield Osborn, *Our Plundered Planet* (Boston: Little, Brown, 1948); Karl Sax, "Population Problems of a New World Order," *Scientific Monthly*, January 1944, pp. 66–71; Frank A. Pearson and Floyd A. Harper, *The World's Hunger* (Ithaca: Cornell University Press, 1945); Guy Irving Burch and Elmer Pendell, *Human Breeding and Survival: Population Roads to Peace or War* (New York: Penguin, 1947); C. Lester Walker, "Too Many People," *Harper's*, February 1948, 98–104; William L. Laurence, "Population Outgrows Food, Scientists Warn the World," *New York Times*, 15 September 1948; Marjorie Van de Water, "More Mouths Than Food," *Science News Letter*, October 18, 1947, pp. 250–251; "Coming: A Hungry 25 Years," *Life*, April 26, 1948, 30–31; H. R. Tolley, "Population and Food Supply," in *Freedom from Want: A Survey of the Possibilities of Meeting the World's Food Needs*, ed. E. E. DeTurk, *Chronica Botanica* 2:4 (1947/48): 217–224; John Donald Black and Maxine Enlow Kiefer, *Future Food and Agriculture Policy: A Program for the Next Ten Years* (New York: McGraw-Hill, 1949), 66–67.

33. Leonard Engel, "Crops by Magic," *Harper's*, March 1948, 278–284; John A. McWethy, "More Food with Power Farming," *Science Digest*, July 1945, 87–90; Robert M. Salter, "World Soil and Fertilizer Resources in Relation to Food Needs," *Science*, May 23, 1947, 533–538; Robert Price Russell, "American Technology for Starved Lands," *Atlantic Monthly*, September 1947, 50–53; Angus McDonald, "Those Long-Haired Scientists," *New Republic*, September 15, 1947, 34–36; William L. Laurence, "Scientists Promise More Food for All," *New York Times*, 28 December 1949, 27.

34. J. Russell Smith, *The World's Food Resources* (New York: Henry Holt, 1919), 329–354; J. D. Bernal, *The Social Function of Science* (London: Geo. Routledge & Sons, 1939), 348; "Foods and Technology of Tomorrow," *Food Industries*, April 1939, 215; "Man-Made Food Possible," *Science News Letter (SNL)*, September 25, 1948, 198.

35. R. L. Meier, "Industrialization of Photosynthesis and Its Social Effects," *Chemical and Engineering News*, October 24, 1949, 3112–3116f; H. A. Spoehr, "Chlorella as a Source of Food," *Proceedings of the American Philosophical Society* 95 (February 1951): 62–67; John S. Burlew, "Current Status of the Large-Scale Culture of Algae," in *Algal Culture: From Laboratory to Pilot Plant*, ed. John S. Burlew (Washington, DC: Carnegie Institution of Washington, 1953), 3–23.

36. Dean Burk, "Vast Energy from Tiny Plants," *Science Digest*, June 1950, 83–85.

37. Burlew, *Algal Culture*, 4; Harold W. Milner, "Algae as Food," *Scientific American*, October 1953, 31; "Food or Fuel from Algae?" *Science Digest*, April 1954, 65–67; Review of *Algal Culture*, *Food Technology*, January 1954, 18. Note that 65 grams of protein was a generous allowance; many nutritionists today recommend a somewhat lower daily protein intake, although most Americans eat far more.

38. Edgar Taschdjian, "Problems of Food Production," *Bulletin of the Atomic Scientists*, August 1951, 211–212.

39. Harrison Brown, *The Challenge of Man's Future* (New York: Viking, 1954), 144–145; William L. Laurence, "Vital Force Found in Plants May Increase World's Food," *New York Times*, Dec. 31, 1949, p. 1. Barbara Ford writes, "Algae experts tend to be optimistic." *Future Food: Alternate Protein for the Year 2000* (New York: William Morrow, 1978), 186.

40. Leslie Velie, "Food Pumped from Pipelines," *Collier's*, December 1948, 9–14.

41. David Orr, "Prices and the Life Exchanged: Costs of the U.S. Food System," in *Earth in Mind* (Washington, DC: Island Press, 1994), 172–184; Marty Strange, *Family Farming: A New Economic Vision* (Lincoln: University of Nebraska, 1988).

42. The literature is so vast on this subject that it is hard to know where to begin. Useful entry points: Cohen, *How Many People Can the Earth Support?* 50–54; Mike Davis, *Late Victorian Holocausts: El Nino Famines and the Making of the Third World* (London: Verso, 2001); Douglas H. Boucher, ed. *The Paradox of Plenty: Hunger in a Bountiful World* (Oakland, CA: Food First, 1999).

43. Roland Barthes, "Towards a Psychosociology of Contemporary Food Consumption," in *Food and Drink in History*, ed. Robert Forster and Orest Ranum (Baltimore, MD: Johns Hopkins University Press, 1979), 166. For recent samplers of food sociology: Alan Beardsworth and Teresa Keil, *Sociology on the Menu* (New York: Routledge, 1997); Steven Mennell, Anne Murcott, and Anneke van Otterloo, *The Sociology of Food: Eating, Diet and Culture* (Thousand Oaks, CA: Sage, 1992); Marjorie L. DeVault, *Feeding the Family: The Social Organization of Caring as Gendered Work* (Chicago, IL: University of Chicago Press, 1991).

44. Lori Valigra, "Recipes for Beyond a Small Planet," MSNBC, February 18, 1998, http://www.msnbc.com:80/news/144882.asp, accessed February 24, 1998. For an extensive discussion of the virtues of social eating, see "Taste, Health, and the Social Meal," a special issue of The Journal of Gastronomy 7 (winter/spring 1993).

45. Rick Weiss, "Moonstruck: Cosmic Cuisine," *Washington Post*, 1 April 1998, E1.

46. On the "chastening" of biotech companies: Daniel Charles, *Lords of the Harvest: Biotech, Big Money, and the Future of Food* (Cambridge: Perseus, 2001). On catering to consumer ambivalence about modernity: Warren Belasco, *Appetite for Change: How the Counterculture Took on the Food Industry* (Ithaca, NY: Cornell, 1993); Donna R. Gabaccia, *We Are What We Eat: Ethnic Food and the Making of Americans* (Cambridge, MA: Harvard University Press, 1998), pp. 1–9, 149–232; Thomas Frank and Matt Weiland, eds., *Commodify Your Dissent: Salvos from 'The Baffler'* (New York: W. W. Norton, 1997); Jennifer Price, "Looking for Nature at the Mall: A Field Guide to the Nature Company," in *Uncommon Ground: Rethinking the Human Place in Nature*, ed. William Cronon (New York: W. W. Norton, 1996), pp. 186–203; David Remick, "The Next Magic Kingdom: Future Perfect," *New Yorker*, October 20, 1997, pp.210–224.

III

Fixing the Environment

7

When Everybody Wins Does the Environment Lose? The Environmental Techno-Fix in Twentieth-Century American Mining

TIMOTHY J. LECAIN

Today, proposals for a technological fix to environmental problems are often met with considerable suspicion and doubt. Modern ecologically informed observers can easily point to past examples of techno-fixes gone wrong, leading many to believe that such supposed solutions often result in further unexpected negative consequences for the environment. As the Greek philosopher Horace observed more than two millennia ago, "Nature, pitchfork it out how you may, keeps tumbling back in on you." In the popular mind, even when environmental techno-fixes do appear to solve problems for various human interest groups, they often appear to do so only by harming the natural world in some new way. Hence the rhetorical question of the essay title: When everybody wins does the environment lose?

The simple answer to this question is often "yes," at least in regard to the messy pollution problems caused by modern industrial mining. As the two case studies in this essay will demonstrate, in the twentieth-century Western mining industry many apparent solutions to problems of air pollution simply involved moving pollutants from one environmental sink to another.[1] Yet it would be foolish to deny that engineers and scientists have also made considerable progress in mitigating the environmental impacts of twentieth-century industrial civilization. In his recent book, *Something*

New Under the Sun (2000), the historian J. R. McNeill offers compelling evidence that advanced industrial nations like the United States, Germany, and Japan, significantly reduced certain forms of air and water pollution, particularly in the post–World War II period.[2] This data would suggest that, contrary to popular perceptions, environmental techno-fixes have indeed solved many environmental problems, and we can be confident that ultimately everybody is winning — including the environment.

This seeming disparity between the optimistic and pessimistic views of environmental technological fixes suggests the need for a more nuanced means of understanding and evaluating how these techno-fixes occur and by what standards we measure their success and failure. Particularly significant in this regard are the engineers and scientists who most often not only developed the techno-fixes, but also created the basic criteria for evaluating their effectiveness. While it is beyond the scope of this essay to examine environmental techno-fixes in a broad array of major industries, two case studies of several important technological fixes invented and adopted in the area of metal mining and smelting may be suggestive of patterns relevant to other polluting industries.

As one of the twentieth century's largest sources of environmental pollutants, the mining industry almost inevitably became a focal point for some of the earliest major conflicts over pollution problems. In particular, competing economic and political interests frequently clashed over the damages caused to local and regional ecosystems by pollution from large-scale mineral smelting operations. Farmers, ranchers, and nearby townspeople complained that smelter smoke damaged their crops, land, and stock animals. In response, smelter operators typically claimed they had little choice but to release these pollutants if their companies were to remain economically viable and produce the minerals that nearly all agreed were essential to industrialization and the health of both state and national economies. To allay these often bitter and protracted conflicts, engineers and scientists were typically called on to develop what we would today call technological fixes — machines or processes that would ideally provide a win-win solution to the problems. While there were a wide variety of different ways in which these historical technological fixes operated, close attention to their means and consequences suggests a rough typology:

1. the transformational techno-fix
2. the relocational techno-fix
3. the delaying techno-fix

Many of the technological fixes to be discussed in this essay consisted of a combination of two or more of these categories, and no doubt further categories or refinements could be developed. Yet these categories provide useful insights into how early pollution problems were solved, and just as importantly, why they were perceived as having been solved. The engineers and mining companies of the day generally viewed the use of any one of these techno-fixes as an appropriate and effective solution to the environmental problem at hand. On further examination, however, all three types of techno-fixes ultimately resulted in other environmental problems, and they thus offer some insight into the strengths and limitations of the engineering and scientific methods used to create and evaluate techno-fixes.

Historians have tended to view the pollution solutions offered by mining and smelting engineers as self-serving measures that simply allowed their employers — the mining companies — to continue to freely exploit the environment by deflecting public criticism.[3] To be sure, mining and smelting engineers were often very closely attuned to the interests of their employers, and they understood that their techno-fixes were a means by which their employers could continue profitable mining operations. Yet to merely dismiss their technological solutions as self-interested expedients tends to obscure the more complex nature of the engineering techno-fix. This is particular apparent when it is recognized that many, if not the majority, of these engineers and scientists clearly held deeply affectionate and respectful views of the natural world, and that both their rhetoric and actions suggest they were genuinely attempting to mitigate the environmental damages caused by mining and smelting. Yet if this was the case, why did the engineers and scientists adopt the limited environmental techno-fixes that they did, and why did they largely ignore their secondary environmental consequences? The answer, at least in part, is that the mining engineers tended to view both the industrial and natural world in terms of their technological knowledge and abilities. Environmental issues were studied and understood only insofar as they affected the operation of complex technological systems.

The evidence that mining engineers valued various types of natural environments is abundant, though it can only be treated briefly here. As early as the post–Civil War years, when the profession first began to develop in the United States, mining engineers frequently expressed their deep appreciation for wild, unspoiled nature. A quote from Rossiter Raymond, one of the leading early mining engineers of the late nineteenth century, is both suggestive and representative. Traveling in Utah in the late 1880s, Raymond was struck by the sight of the Wasatch Range rising from a sea of snow, writing ecstatically of "the orange clouds where the sun had set, ... an exquisite rose-tinted Abendroth on the upper half of the Wasatch Peaks, and over them in the green sky a silver, *strictly silver*, moon!"[4] While the prose of every mining engineer might not have been so purple, many shared Raymond's fondness for a western landscape that they viewed as largely untouched by civilization and the machine.[5]

Yet at the same time Raymond and other engineers were admiring the beauty of western peaks, they were busily designing mines and smelters to exploit the mineral wealth buried beneath them. Islands of advanced industrial technology in a sea of relatively pristine nature, these mining complexes initially seemed environmentally insignificant. While the mining and smelting processes indisputably scarred the landscape and damaged the environment immediately around the mine, most engineers did not find this particularly troubling, in part because the scale of the impact appeared so limited. But by the early twentieth century, the broad environmental effects of smelter smoke in particular gradually began to force mining engineers to develop new technological solutions to environmental pollution.

One of the earliest and most important examples of a pathbreaking technological fix occurred in the southern Smoky Mountains near the junction of Tennessee, North Carolina, and Georgia. In the late nineteenth century, prospectors in the region discovered a large deposit of rich copper ore that rapidly became a center of concentrated mining activity. From 1890 to 1904, operators in the district processed this high-sulfur copper ore by *heap roasting*, an ancient technique in which the smelter men constructed large piles of copper ore and wood out in the open. Once set on fire, these heaps smoldered slowly for days, flooding the surrounding countryside with clouds of sulfur dioxide gas. Within a few years the sulfur dioxide–laden smoke from the heaps transformed what had once been a "beautiful, mountainous, and heavily wooded" landscape into a barren wasteland. Witnesses reported that within a radius of several miles from the Ducktown smelters scarcely a single blade of grass grew, and a wide swath of dead or damaged broadleaf trees could be easily followed for at least thirty miles out into the forests.[5] (See Figures 7.1 and 7.2.)

In the 1890s farmers and townspeople in the Ducktown area began to sue the smelter operators, often winning damage payments, but failing to stop the operators from releasing toxic smoke. Eventually the United States Supreme Court granted the state of Georgia an injunction halting all smelting in the Ducktown district. However, instead of shutting the smelters down, Georgia officials agreed to allow the smelters to continue operations provided the operators removed most of the sulfur dioxide from the smoke. Under pressure from this uncompromising demand either to clean up or shut down, in 1908 the Tennessee Copper Company engineers and chemists developed the seemingly perfect techno-fix: a recovery system that converted the sulfur dioxide gas into sulfuric acid.[6] (See Figure 7.3.) The sulfuric acid plant not only substantially reduced the amount of sulfur dioxide released into the air, but it also established a profitable by-product industry for the smelter operators who found a ready market for their sulfuric acid with nearby manufacturers of superphosphate fertilizers. As one historian of the case concludes, "the site actually represented one of industry's early successes in rectifying — given the standards of the time — a serious and legally complicated interstate air pollution problem."[7]

At first glance, then, the Ducktown story seems to be an example of a completely successful technological fix. The sulfuric acid technology developed by the engineers produced a solution that was reasonably

Fig. 7.1 "A train carries copper ore to the smelter in Ducktown, Tennessee, circa 1939. Even three decades after the sulfur dioxide emissions from the smelter were greatly reduced, the landscape surrounding the smelter remained denuded of almost all vegetation." (Library of Congress, Prints & Photography Division, FSA-OWI Collection, Reproduction Number LC-USF34-052169-D DLC.)

acceptable to all the disputing parties: the agricultural and timber interests were protected from sulfur dioxide, and the smelter operators could continue operating profitably. But while the technology offered a win-win situation for the humans and their immediate environment, it is less clear that the environment as a whole emerged a winner. Instead, the Ducktown case offers a clear example of the first and second types of solutions: The transformational and relocational techno-fixes.

First, the Ducktown smelter operators used the new technology to capture and transform the sulfur dioxide into sulfuric acid. This transformation of a noxious waste gas into a useful product then allowed them to sell the erstwhile air pollutant to fertilizer manufacturers, who further transformed it into superphosphate fertilizer.[8] Having gone through two stages of transformation, the fertilizer makers offered the new product for sale, which ultimately brought about the second techno-fix: the relocation of the sulfur onto farmers' fields across North America and Europe. Indeed, by allowing farmers to replace depleted reserves of phosphorous in intensively farmed soils, the superphosphate fertilizer, and hence the sulfur dioxide, played a key role in increasing global food supplies during the twentieth century.[9]

Having classified the techno-fixes in this way, it is clear that the transformation and relocation of the sulfur compounds did provide an effective

Fig. 7.2 "Billboards along the highway into Ducktown dominate a barren countryside stripped of topsoil and scarred by erosion. The Abernathy Furniture Co.'s sign notwithstanding, the town's forbidding moonscape could hardly have promised much of a 'home sweet home.'" (Library of Congress, Prints & Photography Division, FSA-OWI Collection, Reproduction Number LC-USF34-052170-D DLC.)

solution to the conflicts between farmers and smelters in the Ducktown region. However, even new products as seemingly beneficial as phosphate fertilizers were not without their negative environmental consequences. In order to insure that an adequate amount of fertilizer actually reached the roots of their crops, farmers applied more superphosphate to their land than was actually taken up by the plants. The majority of the fertilizer thus eventually ran off into local creeks and rivers, and in many regions often ended up concentrating in the water of nearby ponds and lakes. In some cases, these high levels of superphosphate led to eutrophication, a process where fertilizer-fed algae and bacteria overwhelm lakes and ponds, absorbing much of the oxygen in the water and choking off most other aquatic life. Wherever phosphate and other fertilizers have been widely used, considerable eutrophication of water systems has resulted.[10]

The engineers and scientists who designed the Ducktown techno-fix, as well as their professional colleagues, appear to have given little or no thought to these secondary environmental effects. Several decades later the prominent physical chemist and smelter expert Robert Swain offered a typically laudatory view, arguing that the Ducktown techno-fix was simply one of the "great industrial achievements of this country."[11] Yet Swain was hardly a man who cared nothing for the natural environment or who would favor its reckless destruction and pollution. To the contrary, like

Fig. 7.3 "A Model example of the transformative techno-fix, the Duck-town sulfuric acid plant removed a large part of the sulfur dioxide fumes produced in smelting copper and transformed it into a new product that was valuable for manufacturing fertilizers and many other industrial processes." (Library of Congress, Prints & Photography Division, FSA-OWI Collection, Reproduction Number LC-USF34-052180-E DLC.)

many other engineers and scientists involved with the smelting problem, Swain was an avid outdoorsman, spending many of his weekends hunting, fishing, and hiking in the California mountains near Palo Alto. Likewise, Swain's career and writings suggest a genuine belief that scientists and engineers should and could develop effective techno-fixes to environmental problems.[12] Clearly, merely suggesting that Swain and other engineers did not care about the secondary environmental consequences of techno-fixes is an inadequate explanation. Rather, a good case can be made that the very nature of the transformational and relocational techno-fixes tended to encourage Swain and his colleagues to view them as environmental successes.

The engineers and scientists who studied the Ducktown problem (and similar smelter smoke problems elsewhere) developed pathbreaking new knowledge about the effects of sulfur dioxide fumes on trees and other vegetation.[13] However, these experts had far less knowledge about the environmental effects of the secondary products the captured sulfur dioxide was transformed into: sulfuric acid and fertilizer. Likewise, the engineers and scientists had little specific knowledge about the new environments the waste products were relocated to: agricultural fields and surface water systems. In the post-World War II period, other scientists developed a clearer concept of the secondary environmental effects of capturing and utilizing sulfur dioxide for superphosphate fertilizers. But in their first tentative efforts to understand and mitigate the environmental effects of smelter pollutants, smelter smoke experts inevitably viewed the environment through the prism of the specific problem at hand and their own technological abilities to solve it. Their research had demonstrated the damaging effects of sulfur dioxide smoke on vegetation; but the transformational techno-fix turned sulfur dioxide into new products whose environmental consequences they had not yet investigated and had little reason to ever investigate. Likewise, they developed significant knowledge about what the smoke did to the ecosystem around Ducktown; but the relocational techno-fix removed the pollutants to new ecosystems they knew little or nothing about. From this perspective, it is clearly a mistake to consider the developers of the Ducktown techno-fix as simply unconcerned or ignorant about environmental matters. Rather, their particular technological solutions encouraged them to develop significant and often pathbreaking knowledge about the effects of one particular pollutant in one particular ecosystem. In other words, the technical experts developed environmental knowledge only insofar as it was directly relevant to the technological systems they were attempting to improve. Once the pollutant was altered into a new product and shipped elsewhere for use, it moved well beyond their still-limited ecological attention and understanding. Thus, by the environmental and technological criteria the engineers and scientists had themselves developed, the pollution problems were indeed solved.[14]

Another common problem with copper smelting in many regions has been arsenic pollution, and as with the sulfur dioxide problem, various engineers and scientists developed powerful new technological fixes to try and solve it.[15] One of the most successful of these was an early-twentieth-century technology invented by Frederick G. Cottrell, a young professor of physical chemistry at the University of California, Berkeley. Cottrell's device, which he called an electrostatic precipitator, quickly demonstrated its potential by removing arsenical particulate pollution from copper smelters in California and Utah. Soon after, his invention attracted the attention of the giant Anaconda Company of Montana, which was facing serious legal challenges to its smelting operation due, in part, to arsenic pollution. In the early twentieth century, Anaconda erected the massive Washoe smelter in the Deer Lodge Valley of Montana. With a price tag of $9.5 million, the Washoe complex was the largest and most efficient copper smelter in the world. But despite the history of litigation over pollution from other copper smelters, the Anaconda Company initially made almost no attempt to control smelter emissions. Rather, the company counted on the smelter's rural isolation and their outright ownership of the nearby town of Anaconda to deter complaints.[16]

The Anaconda Company seriously misjudged the matter. Within months of firing the smelter furnaces in January of 1902, valley farmers and ranchers reported unusually high livestock deaths.[17] By November thousands of horses and cows died in the valley. The ranchers and farmers of the Deer Lodge Valley had little doubt about what killed their animals. As one farmer noted, the new smelter "belches forth such an enormous volume of … poisonous gases that the mind is incapable of comprehending the vast polution [sic] of the atmosphere in its vicinity." Autopsies performed by state veterinarians soon confirmed their suspicions, revealing that the animals had ingested deadly amounts of arsenic.[18] Experts later estimated that the smelter released some twenty tons of arsenic over the farms of the Deer Lodge Valley every day.[19] (See Figure 7.4.)

In 1907 a group of the Deer Lodge Valley farmers and ranchers entered into a protracted legal battle with the Anaconda Company that was eventually resolved in the company's favor. Meanwhile, though, the federal government had also become interested in the Anaconda smelter smoke problem. Environmental investigations made under the direction of President Theodore Roosevelt showed conclusively that the smelter smoke was damaging trees in the nearby Deer Lodge National Forest. Under the

Fig. 7.4 "The giant smokestack of the Anaconda Company's Washoe copper smelter looms over the Deer Lodge Valley in southwestern Montana. As at Ducktown, the sulfur dioxide fumes destroyed vegetation for miles around, but here the smoke also carries large amounts of arsenic that poisoned and often killed livestock." (Library of Congress, Prints & Photography Division, FSA-OWI Collection, Reproduction Number LC-USF34-027562-D DLC.)

threat of a federal lawsuit, the Anaconda Company agreed to allow a board of engineering experts to make a thorough investigation of the problem and to adopt whatever pollution controls it deemed necessary. This Anaconda Smoke Commission had three members: Joseph Holmes, director of the U.S. Bureau of Mines, represented the federal government; John Hays Hammond, a well-known mining engineer and manager, served as an independent technical expert; and Louis Ricketts, manager of an Anaconda-owned mine in Mexico, represented the interests of the Anaconda Company.[20] Given its membership, the Anaconda Smoke Commission unsurprisingly had a strong pro-mining slant, and the members tended to support the Anaconda's desire that any viable solution be an attractively profitable business proposition, similar to that of the Ducktown sulfuric acid operations.[21]

To head up the groundbreaking smoke- and environmental-research efforts at the Anaconda smelter, the Smoke Commission appointed the Bureau of Mines' Chief of Physical Chemistry, Frederick Cottrell. As the director of the research program for the Smoke Commission, Cottrell's work at his small laboratory near the Anaconda smelter would provide the basic technical and environmental information needed for the commissioners to determine what actions the Anaconda Company had to take to alleviate the smoke problem. Simultaneously, both federal and company investigators studied the environmental effects of the arsenic on the surrounding ecosystems, providing data that supplemented Cottrell's focus on the arsenic's complex transition from the smelting process out into the atmosphere. Confident that much of the arsenic pollution could be captured with his electrostatic precipitator, Cottrell made detailed measurements of the pollutants at every stage of the roasting and smelting process and then installed a small experimental precipitator in November of 1913. Encouraged by Cottrell's initial results, the Anaconda Company agreed to purchase the rights to use the Cottrell process from an independent, nonprofit company Cottrell had created to encourage the widespread use of his new technology.[22]

Concerned about the profitability of capturing the arsenic, the Anaconda Company delayed installation of a full system of electrostatic precipitators until 1923. But once installed, the precipitators performed well, and the commissioners were able to report that the Washoe arsenic output was less than one-third of its previous level. As to the third that remained (still as much as twenty-five tons of arsenic per day) the commissioners confidently asserted it was of no further "nuisance to the outside surrounding community" and deemed the problem solved.[23] But having finally been obliged to invest more than $1.6 million to capture arsenic, the Anaconda Company quickly began an intensive search to find profitable markets for a substance that for decades had poisoned the horses, cattle, and people of the Deer Lodge Valley. In keeping with the Ducktown example, the Anaconda initially turned to the agricultural industry as a potential buyer. They discovered that some manufacturers used arsenic to make an insecticide for killing cotton boll weevils in the South, and they soon began selling a fair amount of Anaconda's arsenic for that purpose. Shipping costs, however, cut into the profitability of the pesticide market, so the company also developed an in-house use of the arsenic as a preservative for mine timbers. By 1933, the Anaconda's arsenic plant was producing about 14,000 tons of arsenic per year for these applications, and it realized a net profit of $111,867.[24] (See Figure 7.5.)

Fig. 7.5 "The Anaconda arsenic plant at the base of the Washoe's towering 585-foot smokestack. The plant depended on the powerful techno-fix offered by Frederick Cottrell's electrostatic precipitator to remove a large part of the arsenic from the smoke stream. By 1933, the plant was producing 14,000 tons of arsenic per year, much of which was subsequently used in manufacturing pesticides and timber preservatives." (Courtesy of Montana Historical Society, Helena.)

As with the sulfur dioxide technological fix at Ducktown, the Anaconda solution to the problem of arsenic air pollution looks promising until the full dimensions of the technological fix are analyzed. Indeed, the Anaconda solution illustrates all three types of techno-fixes. The initial profitable use of large amounts of the arsenic involved its transformation into a pesticide and its relocation to southern cotton fields. The precise environmental effects of these arsenical fertilizers are difficult to determine, but we can be sure they were not altogether benign. Arsenic is an elemental substance and natural processes never break it down or alter it into a less biologically harmful form. The poison thus continues to accumulate in soil, water, and organisms over long periods. Obviously, the substance is highly toxic to humans and many other animals when ingested in high concentrations — concentrations that would rarely result in agricultural uses when the pesticide was properly applied. But many scientists also now believe that even relatively low levels of arsenic may cause cancer or other serious diseases in humans and other animals. Likewise, the use of pesticides of any type has increasingly come under fire by scientists who argue that they disrupt ecological balances that naturally keep insect populations in check and can lead to pesticide resistant super-insects.[26]

The ultimate fate of the arsenic that the Anaconda Company stored on site and used as a wood preservative also offers a good example of the third category of solutions, the delaying techno-fix. When the Washoe smelter finally shut down in 1980, the Anaconda Company left behind more than 250,000 cubic yards of dust captured from the smelter smoke. The very ability of the precipitators and dust chambers to winnow out the arsenic from the other elements in the smoke actually helped to increase the arsenic concentrations of this dust to deadly levels of 59,900 to 69,600 micrograms per gram. This arsenic pollution was neither transformed nor fully relocated; rather, the environmental effects of the arsenic were simply delayed. Unlike the original effects of the arsenic air pollution, which were clearly and immediately observable, the effects of the captured arsenic were more gradual and insidious. Over the course of decades, small amounts of the arsenic were spread around the property near the smelter by wind and water. In the early 1980s, state health investigators belatedly discovered that the children in the nearby town of Mill Creek had dangerously high levels of arsenic in their urine. In 1987, unable to make the area safe, the EPA evacuated and relocated all the residents of the Mill Creek community.[27]

Finally, there is the techno-fix in which the Anaconda pressure-treated mine timbers with an arsenical solution to help preserve them in the hot and humid underground mines. While this new use involved both transformational and relocational techno-fixes, it also entailed a delaying techno-fix. For decades, much of the arsenic that had once polluted the air of the Deer Lodge Valley was effectively returned to the underground environment it had initially come from. During most of this time it posed little environmental threat once the miners had installed the timbers and ceased handling them. Yet the benefits of the technological fix did not endure. In 1980 when the underground mines at Anaconda were finally closed, engineers shut down pumps that had removed groundwater from the mine for decades. As the water resumed its natural level, it flooded hundreds of miles of underground passages lined with the arsenic-treated timbers. Thus, decades after it had been removed from the smelter smoke stream, the arsenic that would have otherwise polluted the air over the Deer Lodge Valley is instead now leaching into the groundwater beneath the town of Butte, helping to render it unfit for human consumption or irrigation.[28]

As with the Ducktown case, the success of the technological fixes used to solve the arsenic pollution problem at Anaconda is finally ambiguous. Transforming, relocating, and delaying the effects of smelter smoke arsenic eliminated a pressing local environmental danger. Further, it allowed the two competing interests — the farmers and the Anaconda Company — to continue to both work in the valley for more than half a century.[29] But on a broader geographical and temporal scale, the techno-fix solution was to at least some degree illusory. The problems of that time and that place were met largely by shifting them to future generations and other places.

However, as with the Ducktown problem, the ultimate failure of the Anaconda techno-fixes should not suggest that the engineers and scientists who developed this solution were merely the uncaring servants of a careless profit-minded corporation. To the contrary, Frederick Cottrell, whose revolutionary electrostatic precipitator was fundamental to the technological solution of the Anaconda problem, was a man who cared a great deal about the natural world and dedicated much of his professional life to solving the problem of air pollution. During his childhood in Oakland, California, Cottrell developed an enduring affection for botany and made frequent expeditions into the nearby Piedmont Hills in search of specimens, toting a two-foot long japanned tin case of his own design to carry the day's collection of botanical specimens on the way home.[30] As an undergraduate at the nearby University of California, Berkeley, Cottrell developed a passion for tramping, and he made frequent trips into the Sierras, particularly favoring Yosemite and King's Canyon. His friends knew him to be a powerful hiker, who liked to take giant strides up steep mountain trails, and he remained an avid camper and mountaineer his entire life.[31] This lifelong passion for nature was also reflected in Cottrell's professional career as a physical chemist and inventor. Indeed, Cottrell clearly developed his electrostatic precipitator out of a desire to solve the many environmental and public health problems resulting from large-scale mineral smelting. His selfless dedication to this cause was further demonstrated by his 1912 decision to transfer his precipitator patents to a nonprofit corporation, thus depriving himself of the huge profits from what eventually turned out to be a vastly successful invention. To Cottrell, the goal of solving pollution problems outweighed even his own economic interests.[32]

Like Swain and the many other scientists and engineers who worked to solve the problems of smelter smoke pollution, Cottrell clearly possessed a genuine desire to mitigate the environmental damages caused by smelting. Indeed, his basic research was essential to the development of an early industrial ecology of the smelting industry that helped to explain and mitigate the consequences of arsenic pollution in the Deer Lodge Valley and other smelter sites around the nation. However, there is no evidence to suggest Cottrell or the other experts who worked on the arsenic problem gave any serious thought to the secondary consequences of the captured and reused arsenic. As with the Ducktown techno-fix, the transformation and relocation processes took the arsenic outside of Cottrell's narrowly focused areas of environmental expertise and knowledge. Further, in the Anaconda case study, the delaying techno-fix helped postpone any need for Cottrell or other experts to reckon with the problem of what to do with the captured arsenic for many decades to come. At the very least, the delaying techno-fix offered a short-term solution to an immediate and pressing technological and environmental problem. In the end, Cottrell's techno-fix allowed Anaconda's technological system to coexist with at least one narrowly defined part of the environment.

The Ducktown and Anaconda case studies thus suggest we should regard environmental techno-fixes with some caution. The modern prejudice against such technological solutions appears to be, at least in part, historically justified: environmental improvements in one area have clearly resulted in increased degradation in other areas or in the postponement of the damages to later eras. But perhaps by classifying technological fixes in this way, we can move beyond the early scientific and engineering solutions that focused too narrowly on the local and short-term problems of a technological system. Further, we may better grasp what constitutes a true solution to dealing with vexing environmental hazards such as sulfur dioxide and arsenic. In the cases studied here, transformational, relocational, and delaying techno-fixes were thought to offer genuine solutions at the time, particularly by the experts who invented them and established the criteria for evaluating their success. Further, these solutions required scientists and engineers to develop important, if limited, knowledge about interactions between technological and environmental systems. But in retrospect the techno-fixes can be seen to have often disguised the full magnitude of the environmental problems and thus served to justify the continued operations of the smelter industry. In the final analysis, this may be the most damaging consequence of such technological fixes. Rather than face the difficult reality that mining and smelting the copper ores produced substantial amounts of dangerous chemicals that were very difficult to safely use or dispose of, smelter managers used techno-fixes like sulfuric acid plants and electrostatic precipitators to preserve, rather than abandon or fundamentally modify, their hazardous industrial system. Tellingly, because of the environmental dangers posed by arsenic, most modern U.S. mine and smelter operators attempt to avoid exploiting ores containing arsenic whenever possible.[33] Likewise, before the sulfuric acid solution was discovered, a number of copper smelters around the nation simply shut down or refused to process high-sulfur copper ore. In this light, we might well ask how much sooner would these deeper changes have occurred within the industry if the technological fixes had not been available? In many cases, however, simply avoiding certain ores or not mining and smelting at all may not have been a realistic option. In this light, the techno-fixes offered by the twentieth century mining engineers and scientists were at least a tentative step toward the modern goal of harmonizing the divergent demands of technical and environmental systems.

Notes

1. For urban pollutants, Joel Tarr observes similar patterns in his pathbreaking work, *The Search for the Ultimate Sink: Urban Pollution in Historical Perspective* (Akron, OH: University of Akron Press, 1996).

2. J. R. McNeil, *Something New Under the Sun: An Environmental History of the Twentieth-Century World* (New York: W. W. Norton & Company, Inc., 2000): 50–147. Despite these examples of progress on some types of environmental problems, however, McNeil's overall view of what he identifies as a "gigantic, uncontrolled experiment on the earth" is far from sanguine. By contrast, the controversial work of the Danish political scientist Bjorn Lomborg in *The Skeptical Environmentalist: Measuring the Real State of the World* (Boston: Cambridge University Press, 2001) uses statistical analysis of similar data to conclude that the state of the global environment has dramatically improved in recent years and that environmentalists have exaggerated current ecological threats. However, the Danish Committee on Scientific Dishonesty recently concluded Lomborg's work demonstrated "systematic one-sidedness" and was "contrary to the standards of good scientific practice."

3. See, for example, Donald MacMillan, *Smoke Wars: Anaconda Copper, Montana Air Pollution, and the Courts, 1890–1924* (Montana Historical Society, 2001). A compelling history of the smelter pollution disputes in Montana, which will be discussed in further detail later in this essay, MacMillan's account is nonetheless overly simplistic in its sharp condemnation of the motives and tactics of the mining company engineers and scientists.

4. Further examples of mining engineer's expressing their affection for nature can be found in "Nature and the American Mining Engineer," one chapter of the author's dissertation, "Moving Mountains: Technology and the Environment in Western Copper Mining" (Ph.D. diss., University of Delaware, 1998).

5. As with many other Americans, the mining engineers tended to ignore the earlier presence and environmental effects of Native Americans.

6. Charles H. Fulton, *Metallurgical Smoke* (Washington, DC: Bureau of Mines, 1915), 83–84; Robert E. Swain, "Smoke and Fume Investigations," *Industrial and Engineering Chemistry* 41 (1949): 2385.

7. Sulfuric acid (H_2SO_4) is a dense, colorless, oily liquid that is highly corrosive to many organic materials but has many industrial uses. *Encyclopedia Britannica*, 16th ed., s.v. "sulfuric acid."

8. M. L. Quinn, "Industry and Environment in the Appalachian Copper Basin, 1890–1930," *Technology and Culture* 34 (1993): 575–612, quote on 611. Fulton, *Metallurgical Smoke*, 83–84, notes that at least some damage to farm lands continued to occur since mine owners in 1913 paid $16,500 in compensation to area farmers. Smelter operators also cut back production during active crop-growing seasons in order to reduce harmful emissions.

9. Other uses for sulfuric acid include the manufacture of pigments, dyes, drugs, explosives, detergents, and in petroleum refining.

10. McNeil, *Something New*, 22–26.
11. McNeil, *Something New*, 136–137.
12. Quoted in, Quinn, "Industry and Environment," 592.
13. See in particular his overview of the history of smoke control efforts, "Smoke and Fume Investigations." Swain received his Ph.D. in chemistry from Yale and worked as a professor at Stanford University until 1940. See *National Cyclopedia of American Biography*, s.v. "Robert Eckles Swain."
14. See, for example, Fulton, *Metallurgical Smoke*, where much of this knowledge was later collected and published.
15. The concept of technological systems used here borrows heavily from the work of Thomas P. Hughes. See especially his seminal work on electrical power systems, *Networks of Power Electrification in Western Society, 1880–1930*. Baltimore, MD: Johns Hopkins University Press, 1983. Hughes formulates the concept of *reverse salients* to technological systems, problems that impede the efficient operation of the system. Engineers and scientists develop solutions to these problems that then become part of the system as well. In this light, the solutions to the pollution problems posed by smelting can be seen as tentative attempts to incorporate select parts of the natural environment into the technical system.
16. Note that parts of this section on arsenic pollution were adapted from the author's previous work, "The Limits of 'Eco-Efficiency': Arsenic Pollution and the Cottrell Electrical Precipitator in the U.S. Copper Smelting Industry" published in Volume 5(3), (July 2000) issue of *Environmental History. EH* is published jointly by the American Society of Environmental History and the Forest History Society, Durham, NC.
17. Robert George Raymer, *A History of Copper Mining in Montana* (Chicago: The Lewis Publishing Company, 1930), 67; MacMillan, "A History of the Struggle to Abate Air Pollution From Copper Smelters of the Far West, 1885-1933" (Ph.D. diss., University of Montana, 1973), 107–108.
18. Gordon Morris Bakken, "Was There Arsenic in the Air? Anaconda Versus the Farmers of Deer Lodge Valley," *Montana* 41 (1991): 32; Arthur E. Wells, *Report of the Anaconda Smelter Smoke Commission*, October 1, 1920, National Archives, (Record Group 70, Box 278): 1
19. MacMillan, "Struggle to Abate Air Pollution," 108–112, quote 108.
20. Anaconda Minerals Company, *Mill Creek Remedial Investigation Report* (Anaconda, Montana: Anaconda Minerals Company, 1986), 24.
21. The seminal account of the Anaconda smelter pollution and subsequent legal battles is Donald MacMillan, *Smoke Wars: Anaconda Copper, Montana Air Pollution, and the Courts, 1890–1924* (Montana Historical Society, 2001), but also see the excellent history by Frederic L. Quivik, "Smoke and Tailings: An Environmental History of Copper Smelting Technologies in Montana, 1880–1930" (Ph.D. diss., University of Pennsylvania, 1998).
22. Van H. Manning, *Yearbook of the Bureau of Mines 1916*, (Washington, DC: Bureau of Mines, 1917), 153.
23. Frank Cameron, *Cottrell: Samaritan of Science* (New York: Doubleday & Company, Inc., 1952), 178; Swain, "Smoke and Fume," 2385; Wells, *Smelter Smoke*, 25, 53; J. O. Elton, "Arsenic Trioxide from Flue Dust," *Transactions of the American Institute of Mining Engineers* 46 (1913) 690; H. Welch and L. H. Duschak, "The Vapor Pressure of Arsenic Trioxide," *Technical Paper No. 81* (Washington, DC: Bureau of Mines, 1914).
24. Donald MacMillan, "A History of the Struggle to Abate Air Pollution From Copper Smelters of the Far West, 1885-1933" (Ph.D. diss., University of Montana, 1973), 332-349. Even in modern copper smelting plants designed to capture and utilize pollutants, the levels of arsenic and sulphur compounds that continue to be released have often still been dangerously high. See Peter Dorman, "Environmental Protection, Employment, and Profit: The Politics of Public Interest in the Tacoma/Asarco Arsenic Dispute," *Review of Radical Political Economics* 16 (1984): 151–176.
25. "Arsenic Plant Cost, Sales & Profit Statement (1933)," box 64, folder 3, General Office records, Collection No. 169, Anaconda Copper Mining Company Records, Montana Historical Society.

26. Fumio Matsumura, *Toxicology of Insecticides* (New York: Plenum Press, 1985), 92–93; James C. Whorton, in *Before Silent Spring: Pesticides and Public Health in pre-DDT America* (Princeton: Princeton University Press, 1975), 17, observes that arsenical compounds — usually arsenic trioxide — were the most important and widely used of the first inorganic pesticides.

27. The Anaconda Company initially stored about three-quarters of the arsenic captured by the precipitators: see Frederick Laist to Con Kelley, September 22, 1914, box 63, folder 6, Anaconda Copper Mining Records, Montana Historical Society. On heavy metal contamination of the Clark Fork, see Environmental Protection Agency, "Superfund Program Fact Sheet: Anaconda Smelter Site, Anaconda, Montana," June 1985. On Mill Creek, see Joshua Lipton, *Terrestrial Resources Injury Assessment Report* (Helena, Montana: State of Montana Natural Resource Damage Program, 1993), section 2, 2–5, and *Cleaning up Montana: Superfund Accomplishments* (Helena, Montana: Department of Environmental Quality, 1996), 26.

28. Alan S. Newell, *A Brief Historical Overview of Anaconda Copper Mining Company's Principal Mining and Smelting Facilities along Silver Bow and Warm Spring Creeks, Montana* (Missoula, Montana: Montana Historical Research Associates, 1995), 52; Environmental Protection Agency, Proposed Plan: Rocker Timber Framing and Treating Plant Operable Unit (Washington, DC: Environmental Protection Agency, 1995), 4, 6–7; Cleaning Up Montana, 35. It should be noted that the groundwater is also contaminated my many other pollutant sources; the leaching arsenic is only one of many sources that contribute to this toxic soup.

29. However, MacMillan, "Struggle to Abate Air Pollution," suggests that the agricultural productivity of the valley remained diminished compared to the pre-smelter years.

30. Cameron, *Cottrell*, 38.

31. Cameron, *Cottrell*, 53-54; *Encyclopedia of American Biography*, s.v. "Frederick G. Cottrell."

32. Frederick Cottrell, "The Research Corporation, an Experiment in Public Administration of Patent Rights," *Journal of Industrial Engineering Chemistry* 4 (1912): 864-867; James S. Coles, "The Cottrell Legacy: Research Corporation, A Foundation for the Advancement of Science," in *Cottrell Centennial Symposium: Air Pollution and Its Impact on Agriculture* (Turlock, CA: Cal State Associates, 1977), viii-xvii.

33. Lawrence A. Smith, et al., *Remedial Options for Metals Contaminated Sites* (Boca Raton: Lewis Publishers, 1995), 17.

8

Solving Air Pollution Problems Once and for All: The Potential and the Limits of Technological Fixes

FRANK UEKOETTER

In January 1970, California governor Ronald Reagan invented the perfect cure for the problems of atmospheric pollution. Affirming his administration's "continuing commitment to an all-out war against the debauching of our environment," he came up with what he thought was a brilliant idea of how to raise millions of dollars "to finance the war against smog." As could be expected from someone like Ronald Reagan, his proposal was surprisingly simple: the state should authorize the sale of special personalized vehicle license plates and place the proceeds in a special environmental protection fund. "If only two percent of the owners of registered motor vehicles in California apply for the plates, revenue for the fund will amount to nearly $3.8 million the first year alone," Reagan declared. Also, with automobiles being the prime sources of atmospheric pollution in the state of California, this approach would allow car drivers to get involved in the cleaning of the atmosphere. "The plan will provide every motorist with the opportunity to help solve the problem he has helped to create. The personalized plates will serve as a symbol of his concern for improving our environment."[1]

On the background of the contemporary debate about automobile exhausts, it is impossible not to think about this proposal as quintessentially Reaganesque. When Reagan launched his initiative, the automobile companies were under severe and widespread criticism for failing to produce a pollution-free vehicle. Starting in the early 1950s with a conflict between Detroit and smog-plagued Los Angeles, this conflict had grown into a large-scale confrontation that was fought with bitterness on both sides.[2] In other words, while the battle cry among environmentalists was for a clean automobile, Reagan offered a symbol that would transform a dirty car into an environmental icon for a moderate fee — a letter of indulgence in the worst sense of the word. Still, the concept of environmentally friendly vehicle license plates was more than simply a bad idea. Reagan's initiative stood in a long line of proposals that alluded to a popular hope: the dream to solve air pollution problems once and for all. Whenever discussions arose on issues of atmospheric pollution, people did not hesitate to come up with technological solutions that promised a definite end to pollution troubles. The history of air pollution control is also the history of the dream of the universal panacea.

In this article I would like to focus on one episode in the history of air pollution control where concepts of this kind had a particularly strong impact: the discussion of the smoke nuisance in the decades before the First World War. As environmental historians have repeatedly shown in recent years, the coal smoke nuisance was by far the most important air pollution problem of the early 1900s in both Germany and the United States.[3] With coal being the most important source of energy and imperfect combustion being the rule, almost every large city was suffering from a huge pall of smoke and soot that soiled everything from clothes to façades, reduced the amount of sunshine in the city, and threatened the health of its inhabitants. Consequently, a vivid debate arose in both countries about what to do against this nuisance, with several parties having a stake in it. One of them was the civic anti-smoke movement, which formed in many American cities in the late nineteenth century. Consisting of civic associations and special smoke abatement leagues, they called for a massive campaign against the smoke nuisance. To be sure, the civic anti-smoke movement was never a typical cross-section of the affected population. Smoke abatement was predominantly an upper-class project, but since the rest of the population remained more or less inactive, the socially exclusive smoke abatement leagues effectively earned a representational monopoly.[4] A second party in the smoke abatement debates was industry, which together with railroad companies bore the brunt of the smoke abatement movement's criticism. Industry and railroads were by no means the only significant polluters; household furnaces contributed to the urban smoke nuisance, too. However, domestic smoke did not emerge as a significant issue in smoke abatement debates until the 1920s, and effective abatement would have to wait even longer than that.[5]

It quickly became clear that smoke abatement would also require the knowledge of experts for coal combustion. Since the smoke abatement debate never became so intensive that a change of fuel use patterns became more than a theoretical option, the debate inevitably centered on how coal combustion could be improved, and that was possible only through the use of expert knowledge. However, experts in coal combustion were by no means a homogeneous group. Generally speaking, there was a continuum that ranged from university-trained engineers to more or less dubious inventors, and it was an important precondition of the smoke abatement controversy that during the early years, the latter by far outnumbered the former. While fuel combustion was a rather marginal field in the mechanical engineering profession, numerous inventors of smoke-consuming devices were aggressively marketing their products. A widely read book, *Combustion and Smokeless Furnaces* of 1915, declared that there were nearly fifteen hundred United States patents in force for "boiler furnace devices."[6] A journal article of that time went even further, asserting that for twenty years there had been "nearly three patents a week" for smokeless furnaces.[7] Even considering that not every inventor brought his device to market, the producers of smoke-consuming devices were a sizeable and important party in the war against smoke. In theory, this party was the perfect ally for the smoke abatement movement; after all, it was offering a solution to the coal smoke problem. But in practice, these inventors were eyed suspiciously by professional engineers, creating an important rift among the experts in the smoke debate.

Producers of smoke-consuming devices generally stood out through their far-reaching claims about what their devices could do. "A smokeless city guaranteed," ran as an advertisement for "Ideal Smokeless Down Draft Boilers" in 1916. "All fuel waste is stopped," with coal savings ranging "from 20% to 50%," and in addition, the boiler would "never wear out or corrode."[8] An inventor from Newark, New Jersey, came up with similar promises: "I successfully demonstrated that by means of my grate under boilers, I was able not only to produce a clear chimney but by consuming the smoke given off the fire bed increased by 50% the production of steam as well as saved 50% of the amount of coal consumed."[9] A brochure of the American Fuel Saving Company was written in a similarly enthusiastic vein: the device "insures perfect combustion" and prevented smoke, it could be used with "any grade of coal," it could "be applied to boilers of every description without alteration or detriment," it was "safe, simple, effective and lasting," and it would "increase your efficiency from 20 percent to 40 percent" (though for some reason the company's guarantee only extended to a saving of ten percent).[10] "From the stack connected with the boiler fitted with the Murphy Furnace there is at no time any perceptible discharge of smoke," the Murphy Iron Works of Detroit, Michigan, wrote in an advertisement promising a fuel saving of 25 percent. Needless to say, the company wrote that the device could be installed "in any type of boiler and any size unit."[11] "Our Fuel Saver will save from 10% to 40% of the coal, any kind of coal," another proclamation ran.[12] The basic promises were identical in most cases: the device would end smoke once and for all, it would work with any type of boiler and any kind of fuel, and it would always save enormous quantities of fuel.

Some of the ideas were simply a fraud. For example, Henry Obermeyer wrote of "fuel-saving powders," which were "entirely ineffective" but nevertheless sold all across the country and abroad; with a certain degree of shock, Obermeyer noted that "even scientific Germany has fallen for them."[13] However, with many devices it was not an issue of outright fraud but rather of exaggeration. The Parsons System of Combustion, which sought to eliminate smoke by injecting heated air over the fuel bed was a case in point. Though refraining from explicitly presenting its device as a cure-all, the Parsons Engineering Company of Wilmington, Delaware, essentially marketed its product in this way: "After much care and thought and many years of experimentation and perfecting of an apparatus, the Parsons System of Combustion is today the one arrangement whereby [the] three natural laws of combustion are satisfied and are under control," a company presentation read.[14] The company stressed the usefulness of the device independent of individual furnace conditions: "The staff of the Parsons Engineering Company have gained a wide practical experience in the solution of smoke abatement and other problems incident to imperfect combustion, an experience covering various parts of the country, using all grades of coal and types of furnaces, both locomotive and stationary."[15] However, a review of the Parsons System of Combustion by Arthur D. Little of Boston revealed a more sober reality. To be sure, the device "undoubtedly reduced the smoke to a considerable extent," but the maximum reduction achieved was only 57 percent, making the claim that virtually all smoke was done away with somewhat dubious.[16] Also, while the report made clear "that the principle of the Parsons Complete Combustion System is correct," it also pointed to "mechanical difficulties" and "imperfections in the design of the apparatus as it now exists."[17] And as a side note, Arthur D. Little also took the air out of the company's claim that this was the only invention of its kind: "The value of supplying air over the bed of fuel has been realized and is embodied in many inventions relating to the prevention of smoke and increased fuel economy."[18] Thus, the summary included a polite but firm refusal of the far-reaching claims that the company had made: "The results of these tests as a whole were not as good as had been anticipated judging from the reports of previous tests."[19]

This conclusion was emblematic of the results that investigations of smoke-consuming devices by professional engineers typically produced. Of course the special findings differed from case, but the bottom line was usually the same. Many devices could actually work, but not as effectively as their inventors claimed. Many of them could significantly reduce the amount of smoke, but only if the device met the specific requirements of the individual case. In theory, smoke abatement was easy: smokeless combustion required a sufficient amount of air, a proper mixture of air and fuel, and a sufficient temperature — the "three natural laws of combustion" that the Parsons Engineering Company had referred to. "Any fuel may be burned economically and without smoke if it is mixed with the proper amount of air at a proper temperature," a standard formula ran.[20] But in practice, it was a complex task to achieve these conditions. Smoke abatement experts had to consider a large number of factors, ranging from the type of fuel to the power requirements of the factory. Smokeless combustion was possible only when all of these factors were taken into account. On this background, it is clear that the popular promise of producers of smoke-consuming devices that their product would work in every situation was not technologically sound. There was probably a solution for every special case, but not *the one* solution. A device that would work perfectly in one furnace could be entirely ineffective in another, or even counterproductive, increasing rather than lessening the amount of smoke. "No one type of stoker is equally valuable for burning all kinds of coal," a common admonition read.[21] Thus, from the point of view of professional engineers, it would have been desirable if the producers of smokeless devices had specified the conditions for which their products were built. However, most of them did the opposite: they promised a universal panacea for every situation, making producers of smokeless devices extremely unpopular in engineering circles.

Consequently, there is no lack of statements attesting to the engineering community's negative attitude. "In many instances devices which will prevent smoke fail to produce the results expected because the conditions under which they are installed are entirely improper," an engineer wrote in *Harper's Weekly*.[22] In a similar vein, the St. Louis–based Special Committee on Prevention of Smoke declared in 1892 that while there were a number of "meritorious devices" on the market, none of them was of unlimited or unconditional effectiveness: "what their practical limitations are, we are compelled to think that neither those who have invented, sold or used them, know."[23]

But many engineers went further than that, chastising smoke-consuming devices as a matter of principle. Victor Azbe, one of the more prominent anti-smoke experts of the interwar years, wrote in one of his publications, "As far as smoke-consuming devices are concerned, most of them are of no value whatsoever, and are merely dreams of inventors who have no adequate conception of the principles of combustion."[24] "Probably there are few engineers who have not had a fair share of nostrums or wonderful cure-alls offered them to correct the troubles of the power house and boiler room. Possibly at the head of these stand the various contrivances called 'smoke-consumers,'" an article in *Power Plant Engineering* noted.[25] "Next to the perpetual motion crank, comes the man who claims to have invented a smoke consumer," William H. Bryan of Purdue University mocked in *Cassier's Magazine*, adding more soberly a few years later that, "the trouble is that few inventors realize the complications and difficulties of the problem."[26]

Many comments made it clear that there was a considerable degree of anger in engineering circles, as the following statement during a meeting of the Engineers' Society of Western Pennsylvania in 1892 showed: "I am an earnest advocate of anything that will abate this nuisance, but I do say right here that we have been flooded, as other cities have been, with testimonials of smoke saving devices that are not worth the paper they are printed upon."[27] A German engineer took refuge in irony when he summarized his impressions after a visit to the smoke abatement exhibition in London in 1882: with promises about huge savings abounding, a layman would have to conclude "that by combining several of these devices, he would ultimately need very little coal or even no coal at all."[28] And O. P. Hood, chief mechanical engineer of the United States Bureau of Mines, regretfully noted in an address before the Citizens' Smoke Abatement Conference in St. Louis, "Many foolish cure-alls are presented to the public to cure the smoke evil."[29]

It is interesting to note that these comments were not only critical but explicitly aggressive. Apparently, university-trained engineers were not only skeptical toward the inventors' exaggeration, they were angry about them, and for a reason. Fuel combustion enjoyed a rather low reputation compared with other fields of mechanical engineering at the turn of the century; in fact, it was not even clear whether fuel combustion actually *was* an established field of mechanical engineering. In this situation, the exaggerations of producers of smoke-consuming devices were not only a problem of false information, but a full-scale challenge to the engineers' professional ethos. "Considering all the things that have been proposed in this field, it is natural that people show an enormous distrust toward us," a German engineer complained.[30] The inventors' dubious promises threatened to tarnish the reputation of university-trained engineers as well, and in vigorously attacking the producers of smoke-consuming devices, the engineers tried to limit the damage to their profession's reputation as much as possible. Thus, the dispute on smoke-consuming devices was also a conflict about professional jurisdictions.[31]

The engineering profession ultimately prevailed in this struggle, and fuel combustion won a place in the engineering profession during the first half of the twentieth century. However, this victory was never complete: as late as 1938, St. Louis smoke inspector Raymond Tucker bemoaned the "deluge of panaceas" and the "many wild and careless statements" that were being "disseminated about the efficacy of various cure-alls for our present smoke nuisance."[32] So, even after the technological supremacy of university-trained mechanical engineers was firmly established in this field, the producers of smoke-consuming devices continued to market their products as universal panaceas; and that makes this subject interesting for a volume on technological fixes: why did the producers of smoke-consuming devices offer universal panaceas in spite of serious theoretical and practical objections from the engineering community? To be sure, the companies' marketing position was certainly part of the story. Since almost every producer of smokeless devices offered a cure-all, it was difficult for the individual producer to be honest about the limits of its product. After all, how would an invention with specific requirements come across in a market where everyone else offered universal panaceas? However, on second thought, the limits of this explanation become plainly apparent. The producers' exaggerations were bound to produce disappointments, which were almost certain to bring the entire range of products into disrepute. Also, the popular claim that a certain device was the only viable solution automatically discredited all other producers of smoke-consuming devices. For example, when the American Fuel Saving Company claimed that its product yielded "better results than any other device ever offered," this obviously did little to raise the reputation of the field as a whole.[33] Thus, in order to explain the popular promise of the cure-all, it is insufficient to focus only on the producers who offered them and the experts who criticized them. After all, someone had to buy the producers' promises, literally and figuratively. The rhetoric of the universal panacea must have met with certain wishes and desires in the audience. People must have had good motives to ignore the warnings of the experts and listen to the soothing words of the advertisements. And these motives tell us something about popular conceptions of technological fixes.

As mentioned above, the smoke abatement movement focused particularly on industry and railroad companies during the early years, and consequently, the vast majority of smoke-consuming devices were made for these customers. Environmental historians have long argued that industrialists sought to avoid unproductive expenses, making them fierce opponents of smoke abatement and any other kind of environmental regulation. However, a look at the actual behavior of businessmen shows that things were, if anything, more complicated.[34] In the majority of cases where entrepreneurs were on the spot for producing smoke, they did not show a prominent interest in cost reduction, but rather no interest at all. In order to understand the reaction of industry, it is important to realize that smoke from the combustion of coal was about the last thing that a typical industrialist was interested in. "The boilers are generally placed in the very darkest corner, in places the owners seldom care to visit," an expert from Pittsburgh declared in a meeting of the Franklin Institute in 1897.[35] "It is an unfortunate fact that the plant owner or manager knows less about his power plant than he does about any other part of his establishment," Cincinnati smoke inspector Gordon Rowe remarked before the annual convention of the Smoke Prevention Association in 1922.[36] In a treatise written for school use, Harry B. Meller and Logan Sisson of the Mellon Institute of Industrial Research in Pittsburgh wrote, "To so many persons, a boiler is just a boiler, and they think that any one will do so long as it will burn fuel and produce heat."[37] And Raymond Tucker declared in a speech before the National Association of Power Engineers, "It has been the experience of our Division that in the isolated power plants that the boiler room is the 'forgotten man' in engineering."[38]

On this background of widespread lack of interest, it becomes understandable why businessmen so frequently believed the promises of the producers of smoke-consuming devices. If an industrialist received a notice urging him to control his smoke but was not interested in his furnace room, a device that promised to end the trouble once and for all must have appeared like a dream come true. The cure-all rhetoric met with an expectation in the business community that by buying some simple device, the issue would be taken care of. With costs ranging from $200 for a hot-air system to $900 for underfeed or chain stokers, smoke-consuming devices were probably not a bargain.[39] But if that price bought freedom from the hassler in the department of smoke abatement, an industrialist was strongly inclined to invest in a presumed universal panacea. When notified of smoke violations, entrepreneurs simply bought "some cheap and useless device," the Chicago Department of Health reported in 1887.[40] Being more or less ignorant of the essentials of smokeless combustion, businessmen typically had little chance to detect exaggerated promises. If an industrialist did not even know the ratio between coal consumption and steam production, as was frequently the case, he had little chance to disprove the advertised coal savings or to decide which expert he could trust. Few businessmen would know how to distinguish between a dubious inventor and a professional engineer.

Thus, most businessmen were simply hoping for an end to smoke troubles when they bought a smoke-consuming device. The question whether the device met the specific requirements of his plant scarcely received any attention. The inevitable result was a large number of ill-equipped furnaces. Many devices failed to achieve the desired results, much to the chagrin of the business community. "Much disappointment has arisen and much unnecessary expense been incurred by owners of plants not giving sufficient consideration to their own particular requirements before selecting a device," the St. Louis Smoke Abatement Department wrote in a 1909 report.[41] Similarly, Herbert Wilson of the U.S. Bureau of Mines noted, "Many manufacturers who have been looking for some cheap addition to a poorly constructed furnace to make it smokeless have experienced the inevitable failure."[42] When the German city of Karlsruhe conducted a systematic study of the local smoke situation, it found that about 50 percent of the installations found to create a smoke nuisance actually had a smoke-consuming device.[43] The Chicago Department of Health reported that in 1897, the number of repairs made to smoke-consuming devices by far exceeded the number of new devices installed.[44]

With some businessmen, the hope for a cure-all died slowly. For example, the Special Committee on Prevention of Smoke in St. Louis found out that a local businessman had invested $8,000 on unsatisfactory experiments with smoke consumers.[45] "We have used two devices and are now trying a third," a manufacturer from Rochester wrote in response to a questionnaire from the local chamber of commerce, adding that, "none of these have proven entirely effective under all conditions."[46] Obviously, this company was bound for the thoroughly negative opinion on the subject of smoke abatement that many industrialists held as a result of hapless experiments with smoke consumers. "When the owners of plants have spent from $75 to $800 on each boiler for a device that will prevent smoke, they will not believe but that they are the objects of unwarranted and malicious attacks when told that they are still violating the ordinance," Chicago's smoke inspector reported in the 1890s.[47] A few years later, the Illinois Manufacturers Association wrote in a letter to city council members, "Many of those who are skeptical have done much in an experimental way and have spent thousands of dollars in their efforts to stop their chimneys from smoking and have made an honest effort to do their part toward freeing the atmosphere from pollution."[48]

Thus, the net result was that all parties involved lost in the long term: industrialists grew skeptical of smoke abatement in general; civic smoke abatement advocates found their cause being discredited; engineers felt their own professional reputation imperiled; and the producers of smoke consuming devices created a hostile climate among businessmen. "The men who sell devices ... have done more by their mistaken zeal and unreasonable representations to disappoint and discourage our manufacturers in their honest efforts to abate the smoke nuisance than all other influences combined. Of course this short sighted policy presently defeats itself," the president of the Smoke Abatement League of Hamilton County (Cincinnati, Ohio) said in a speech to the National Association of Stationary Engineers in 1906.[49] Similarly, authorities in the industrial town of Mülheim in the vicinity of Cologne wrote that statements of industrialists on smoke-consuming devices were "almost generally derogatory."[50] The ministry of the interior of the state of Baden wrote of "widespread prejudices" concerning these devices.[51] An expert commission in the state of Prussia spoke of a "justified distrust" among industrialists.[52] Businessmen experimenting with smoke-consuming devices soon found themselves caught in a vicious circle of illusions and disappointments.

In both countries the producers of smoke-consuming devices were not only targeting businessmen but also civil servants in charge of smoke abatement. Interestingly, German officials were much more prone to proposals of this kind than their American colleagues. Most German bureaucrats dealing with the smoke nuisance were not trained engineers, making them just as knowledgeable as the average businessman. Therefore, producers of smoke-consuming devices sometimes succeeded in finding an ally in the administration. For example, a public health committee in the state of Württemberg became convinced that a product of the Löffler Company was the ultimate solution of the smoke question. Consequently, the committee sent several letters to other institutions urging them to pay attention to this presumed universal panacea (while the Löffler Company diligently supplied the committee with advertising material).[53] In

Nürnberg, city officials repeatedly forced businessmen to install a Ten-brink furnace, and in the city-state of Bremen, the factory inspector once came to believe that the Donneley smokeless device was the ultimate solution of the problem, converting only after numerous devices had failed in practice.[54] To be sure, stories of this kind were the exception rather than the rule. Nevertheless, they show that the concept of a cure-all against smoke was popular not only among industrialists but also among bureaucrats. The motive was basically the same for both groups: for a person ignorant of the fundamentals of combustion and unwilling or unable to invest much time in the study of smoke abatement literature, the promise of a universal panacea was hard to resist.

So, when producers of smoke-consuming devices were marketing cure-alls, they were trying to tap a tacit hope that many businessmen and some officials held, while otherwise counting on these groups' technological ignorance. From today's perspective, one may wonder whether these producers would have fared better if they had embraced a more honest approach by detailing the limits of their devices; in the long run, this certainly would have brought them more confidence among the purchasers. However, with numerous inventors trying to make a fortune with their proposals (one offered his invention to the city of Pittsburgh for one million dollars![55]) such long-term considerations never had a chance. Consequently, the potential allies of the smoke abatement movement became the target of much criticism; for example, Chicago smoke inspector Viall lamented over the "great many quack device fellows" at a national convention in 1913.[56] "Many good citizens, anxious to comply with the provisions of their local smoke ordinance, become eager purchasers of patent cure-alls, sometimes sold at an extravagant price, the net result being that the entire movement is often discredited," Henry Obermeyer complained.[57] But statements of this kind were obviously insufficient to combat the popular concept of universal panaceas. The smoke abatement movement had to offer a regulatory strategy that would be more attractive than the elusive search for a cure-all. And at this point, the German response differed markedly from the American one.

The main driving force in American smoke abatement discussions were the civic associations concerned about smoke. However, discussions on smoke abatement strategies were always more than sober discussions on ways and means. As the German political scientist Joachim Raschke wrote in his seminal treatise on social movements, "one needs to caution against the assumption that one can interpret a social movement's decisions on strategies as thoroughly rational."[58] At first, the civic anti-smoke movement tried to reach its goal through the vigorous prosecution of smoke offenders. However, it soon became apparent that such a strategy was essentially counterproductive: it brought only moderate success, it required enormous bureaucratic resources, and it antagonized the businessmen that the anti-smoke movement tried (in its own words) to "convert" and "enlighten." Nevertheless, prosecution remained the method of choice for several decades, not so much due to its effectiveness but because this strategy correlated perfectly with the anger of civic anti-smoke activists over smoking businesses. However, the massive side effects of a prosecutorial strategy ultimately forced the anti-smoke movement to reconsider its strategy: instead of prosecution, the movement now favored the education of businesses by professional engineers. Experts in fuel combustion would help companies to correct the deficits of their firing equipment, transforming industry from the object of vigorous prosecution into an ally of the anti-smoke movement. This shift of strategies met with great favor. Ultimately the result was a win-win-scenario: industrialists won a consultant who helped them to abate smoke and save fuel, engineers won jobs in municipal governments, and the anti-smoke movement could rejoice over a significant (though not complete) reduction of smoke emissions.[59]

The only group that could not rejoice over this new strategy was the producers of smoke-consuming devices; but that, of course, was intentional. In fact, the prominence of university-trained engineers in the new strategy against smoke was to a large extent a response to the popular dream of a universal panacea.

While an industrialist by himself was usually helpless when confronted with a salesperson's sweet promises, he could now consult an expert to find out whether the offer was technologically sound. In fact, many smoke inspectors saw it as one of their prime tasks to discipline the producers of smoke-consuming devices. For example, the Chicago Department of Health embraced a policy of giving "very frank advice about installations of so-called smoke devices. The position taken is that these so-called smoke devices in themselves are not sufficient to meet the department's requirements. However, if a plant owner wishes to build a really good furnace and in addition put on the smoke consumer, the department offers no objection." In this way, the department did what the producers of smoke devices had failed to do: it specified the conditions under which smoke-consuming devices would work and directed the businessmen's choice accordingly. This naturally met with great favor among businessmen since this policy allowed them to avoid costly mistakes; the department was firmly convinced that its policy had "saved the public a great deal of money."[60]

Thomas Donnelly, chairman of the Committee on Smoke Abatement of the City Club of Chicago and the spiritual father of the municipal department, heartily agreed. In a presentation before the Smoke Abatement Conference in London in 1912, he even described this policy as one of the key pillars of the department's success: "The engineers of the Department must be experienced and able enough to know just what design will be most effective for any particular condition, and arbitrarily refuse all bad or border-line proposals made by owner, consulting engineer, or the apparatus manufacturer." After all, experience had shown that "the most difficult offender" was the one "who had recently installed a new plant, and who, under incompetent or interested advice, put in what he supposed was a satisfactory smoke preventor, but which had proved a failure."[61]

The new engineering approach to smoke abatement was first inaugurated in Chicago, but it soon ceased to be a local peculiarity. In an article that was written as a guideline for civic reformers trying to define a smoke abatement policy, John O'Connor of the University of Pittsburgh declared that a city "should place special emphasis on securing facts as to the ability of the different installations to abate smoke."[62] In order to deal with the "great many patented furnaces and smokeless devices," O. P. Hood recommended, "You had better call in your mechanical engineer and get some advice before going too far in investment of these."[63] The St. Louis Division of Smoke Regulation noted in its "Report of Operations" for 1937/38 that smoke-consuming devices "should be thoroughly investigated because their use may be a source of annoyance to the user."[64] And in fact this policy seems to have had its effect: the universal panaceas probably did not disappear from the market altogether, but many producers decided to be more honest about the limits of their devices. "Today the leading manufacturers of devices that will abate the smoke will not sell them, except under conditions which they know to be proper," an engineer wrote in 1912, even saying that, "A failure of any of the leading devices for the prevention of smoke is very infrequent nowadays."[65] And in 1927, Chicago smoke inspector Frank Chambers reported, "Manufacturers of stokers and furnaces have cooperated actively by making improvements in their equipment, so that with reasonable care they would be operated more easily in compliance with the smoke ordinance."[66]

Interestingly, a similar institutionalization of engineering expertise did not take place in Germany, at least not on a similar scale. To be sure, there were attempts to set up institutions of this kind, but these failed to gain momentum. The most ambitious attempt was the foundation of the *Hamburger Verein für Feuerungsbetrieb und Rauchbekämpfung,* or Hamburg Society for Fuel Economy and Smoke Abatement, an association founded by industrialists to fend off a pending anti-smoke law. The association hired a number of professional engineers and offered their advice to its members, pointing out that by optimizing the combustion of coal, industrialists would abate smoke and save fuel at the same time.[67] Inevitably, the exaggerated claims of the producers of smoke-consuming devices were one of the challenges that the engineering experts had to meet. In its report for 1906, the society wrote that hapless experiments with smoke-consuming devices "were inclined to bring smoke abatement *per se* into disrepute among industrialists."[68] In an article of 1905, the chief engineer of the association lashed out directly at the producers of smoke-consuming devices: "There is barely a field of engineering where there is a similarly aggressive kind of advertisement with highly exaggerated promises and scores of inventors, showing more than everything else that much remains to be checked and corrected in this field.[69] But like the smoke inspectors in the United States, the engineers of the association ultimately prevailed, at least to a certain extent. In its report for 1910, the society once again mentioned the widespread distrust of industrialists against smoke-consuming devices, but confidently added that the association was "taking the air" out of many exaggerated promises.[70]

Interestingly, the membership of the Hamburg Society for Fuel Economy and Smoke Abatement did not remain confined to the city-state of Hamburg and its environs. Members came from all parts of Germany, showing that there was a general dearth of engineering advice in the field of smoke abatement and fuel combustion. Still, the Hamburg Society remained the only association of its kind inside Germany (although sister associations were founded in Amsterdam in the Netherlands and Helsinki, Finland); the hopeful beginning in Hamburg did not grow into a comprehensive network of professional advisory institutions that was comparable to the American community of smoke inspectors. Instead, the German smoke abatement debate largely centered on the human element and gave firemen great and lasting attention. "A good fireman is the best smoke consumer," a typical and much-heard argument ran.[71] Consequently, German smoke abatement advocates directed a large share of their work toward assuring better training of firemen, while questions of furnace and combustion technology were more or less neglected.

In contrast, American anti-smoke activists rarely mentioned the firemen, and when they did, the firemen appeared as a source of annoyance rather than a person in need of support. "The extent to which the smoke is kept down depends upon the individual efficiency and carefulness of a multitude of poorly paid and, in many instances, ignorant workmen," the Chicago Department of Smoke Inspection wrote with obvious anger in a 1911 report. But where German experts might have argued for increased training efforts, the smoke inspectors proposed to reduce "the personal element … as low as possible."[72] Automatic stokers that guaranteed a continuous stream of coal on the fuel bed became the method of choice. In larger-sized plants, the Chicago department required "automatic stokers unless there is some good engineering reason why they should not be installed."[73] St. Louis smoke inspector Hoffman embraced a similar policy: "I shall endeavor to persuade or force large plants to use automatic stokers," he declared upon his appointment.[74] Other experts agreed, calling hand firing "uneconomical and unbusinesslike" in plants beyond a certain size.[75] "Get away from the human element as far as possible — especially the untrained human element," Pittsburgh smoke inspector Henderson declared.[76] And Herbert Wilson wrote in his essay on "the cure for the smoke evil, … The fireman is so unreliable a factor that the ultimate solution of the problem depends upon the mechanical stoker; in other words, the personal element must be eliminated."[77]

To be sure, firemen were not completely ignored in everyday smoke abatement work. Since even good equipment could produce smoke if poorly handled, the instruction of firemen was an important part of every smoke inspection campaign. In 1914, Chicago chief smoke inspector Osborn Monnett reported that his department was conducting a comprehensive "campaign of education," with ten "expert engineers" spending "most of their time instructing firemen in the proper methods of handling their equipment."[78] Two years later, Monnett mentioned that one of the effects of his systematic smoke abatement work was that businessmen showed more interest in the work of the heretofore neglected firemen: "Many concerns in Chicago saved themselves considerable expense … by giving their firemen a pair of shoes every month if no smoke summonses were received."[79] However, all this never led to a change of policy that would have placed more reliance on the skill of the firemen; the prevalence of the technological fix remained unchallenged. While Germans sought to train firemen until they were skilled enough to burn coal without smoke even under adverse furnace conditions, Americans would confine the efforts to avoiding mistakes that could damage the equipment. With few exceptions, Americans soberly saw the limits of the education of firemen : "One of the very best smoke preventers known is a good fireman, but alas? If he could be any thing else he would not be a fireman."[80] "Foreign countries have given hand firing more attention than we have, particularly industrial Germany, where great stress is laid on the skill of the fireman," the Rochester Chamber of Commerce noted in a brochure on the "smoke shroud."[81] And Imogen Oakley wrote in the *Civic Club Bulletin* of Philadelphia, "In America we have been placing our faith in the various mechanical devices."[82]

In retrospect, it seems like the American approach with its heavy reliance on technological fixes was ultimately more effective than the German one. To be sure, Germans did conduct the training of firemen with diligence and intensity. For example, the city of Aachen employed a special firing instructor who offered on-the-job training to all firemen within the city limits.[83] The Polytechnic Society of Leipzig gave out bonuses to firemen on the basis of systematic smoke observations, and the Hamburg Society for Fuel Economy and Smoke Abatement maintained a similar system.[84] The factory inspector of Bremen took the initiative of creating a firing school, with funding provided by a number of local businessmen.[85] In Königsberg and Hannover, local trade associations ran firing schools, while private associations for boiler inspection (*Dampfkessel-Überwachungsvereine*) maintained such schools in Danzig, Stettin, Magdeburg, and Kassel.[86] As early as 1879, there were firing schools in four cities in the state of Saxony.[87] Since 1902 the Prussian Ministry of Commerce paid an engineer and a firing instructor to run two-week-long courses for firemen in all parts of the state.[88] Between 1906 and 1913, the Prussian state also paid for a second pair of instructors.[89] Finally, the state of Prussia made sure that in all of Prussia, firing instructors were available for the training of individual firemen at the workplace.[90]

However, the net result of all these activities was limited, as a look at the attendance figures shows. In 1910 the Prussian firing instructors visited a total of 8,140 steam plants, while the total number of boilers in the state of Prussia was 107,160.[91] In a memorandum of 1913, the Prussian Ministry of Commerce wrote that the vast majority of firemen still received only rudimentary training for their job, a state of affairs that the ministry was obviously anything but pleased to concede.[92] But even if there had been further options for training, one may doubt that it would have made much of a difference. Many observers stressed with some justification that proper firing was first and foremost a question of character. For example, an official from Dresden wrote that firemen needed skill, reliability, and good will.[93] A good training program could probably improve the skills of a worker, but the two latter factors were obviously beyond the capabilities of an evening school.[94] Finally, there was a good deal of fluctuation among the firemen; many firemen fled into other and better paid jobs as soon as they could.[95] In 1918 the Hamburg Society for Fuel Economy and Smoke Abatement wrote in its annual report that the firemen had remained "a little noticed and little esteemed occupational group," showing how little the training efforts of previous years had effected.[96] The training programs were probably not completely useless, but compared with the merits of the American approach centering on inspections and engineering advice, it seems that in stressing the need of well-instructed firemen, the German smoke abatement advocates ultimately found itself tilting at windmills. Harsh as it may sound, a greater reliance on technological fixes and less compassion for the untrained firemen could have brought Germany a better smoke abatement strategy.

In theory, there was a very simple answer to the popular hope to solve air pollution problems once and for all: there is no universal panacea. But the history of such universal panaceas in the field of air pollution control shows that this response, true as it was, was missing the point. People did not simply believe in universal panaceas because the producers of smoke-consuming devices gave them false information. The cure-alls only found their purchasers because their promise met with a tacit hope to solve the problem once and for all. For a person who had better things to do than read the smoke literature, it was the perfect reaction on first glance to buy and install a certain device and henceforth care no further. Consequently, many people found themselves caught in a vicious circle of high hopes, artificially created illusions, and their subsequent disappointment. University-trained engineers fought hard to distinguish themselves from the producers of smoke consumers and claimed that the subject belonged in their own hands, and they ultimately succeeded in gaining professional jurisdiction over this field of technology. To be sure, their success was never complete because producers of cure-alls stayed in the market and still found industrialists willing to buy their promises (and their devices), but in general, professional engineers came to be seen as the authoritative source of information on the subject. By hiring university-trained engineers as municipal employees and smoke inspectors, businessmen gained access to information that they had been lacking before, thus breaking the vicious circle of illusions and disappointments to a certain extent. The idea was that as soon as businessmen gained access to professional know-how (if only by calling in the municipal engineer), the producers of smoke-consuming devices needed to be more honest about the limits of their products. By focusing on the fireman, rather than the technology of fuel combustion, the German reformers failed to induce a similar process. In stressing technological fixes and disregarding alternative approaches, the American smoke abatement movement followed a path of reform that was ultimately more effective than its German counterpart. This article underscores the importance of a differentiated understanding of technological fixes, and especially the need of a close look at potential alternatives.

Notes

1. Journal of the Assembly, Legislature of the State of California, 1970 Regular Session, vol. 1, pp. 308, 311.
2. Cf. James E. Krier and Edmund Ursin, *Pollution and Policy. A Case Essay on California and Federal Experience with Motor Vehicle Air Pollution 1940–1975* (Berkeley et al., 1977).
3. Cf. R. Dale Grinder, "The Battle for Clean Air: The Smoke Problem in Post–Civil War America," Martin V. Melosi (ed.), *Pollution and Reform in American Cities 1870–1930* (Austin and London, 1980), pp. 83–103; David Stradling, *Smokestacks and Progressives. Environmentalists, Engineers, and Air Quality in America, 1881–1951* (Baltimore and London, 1999); Harold L. Platt, "Invisible Gases: Smoke, Gender, and the Redefinition of Environmental Policy in Chicago, 1900–1920," *Planning Perspectives* 10 (1995), pp. 67–97; Gerd Spelsberg, *Rauchplage. Zur Geschichte der Luftverschmutzung* (Cologne, 1988); Arne Andersen and Franz-Josef Brüggemeier, "Gase, Rauch und Saurer Regen," Franz-Josef Brüggemeier and Thomas Rommelspacher (eds.), *Besiegte Natur. Geschichte der Umwelt im 19. und 20. Jahrhundert* (2nd edition, Munich, 1989), pp. 64–85; Arne Andersen, "Die Rauchplage im deutschen Kaiserreich als Beispiel einer versuchten Umweltbewältigung," Gerhard Jaritz and Verena Winiwarter (eds.), *Umweltbewältigung. Die historische Perspektive* (Bielefeld, 1994), pp. 99–129; Franz-Josef Brüggemeier, *Das unendliche Meer der Lüfte. Luftverschmutzung, Industrialisierung und Risikodebatten im 19. Jahrhundert* (Essen, 1996).
4. On these more general aspects cf. Frank Uekoetter, *Von der Rauchplage zur ökologischen Revolution. Eine Geschichte der Luftverschmutzung in Deutschland und den USA 1880–1970* (Essen, 2003).
5. On the smoke abatement campaigns in St. Louis and Pittsburgh, where domestic smoke was first targeted successfully, see Joel A. Tarr, The Search for the Ultimate Sink. Urban Pollution in Historical Perspective (Akron, 1996), pp. 227–261, and Joel A. Tarr and Carl Zimring, "The Struggle for Smoke Control in St. Louis. Achievement and Emulation," Andrew Hurley (ed.), *Common Fields. An Environmental History of St. Louis* (St. Louis, 1997), pp. 199–220.
6. Jos. W. Hays, *Combustion and Smokeless Furnaces* (Chicago, 1915), p. IV.

7. Frank C. Perkins, "An Electric Smoke Recorder," *Industrial World* 47 (1913), pp. 142–144; p. 144.

8. *St. Louis Globe-Democrat*, 15 Jan. 1916, p. 3, c. 4–7.

9. Archives of the Industrial Society, University of Pittsburgh 83:7 Series I FF 17, Martin to The Smoke Inspector, Smoke Investigation Board, Mellon Institute, September 12, 1913.

10. Hagley Library, Wilmington, DE, Pamphlet File, American Fuel Saving Company, Fuel Saved, Smoke Prevented (Philadelphia, n.d. [1899?]).

11. *Industrial World* 48 (1914), p. 134.

12. Western History/ Genealogy Department, Denver Public Library, Jay T. Williams Papers FF 1, Patee to Mr. Coal Dealer, January 11, 1928.

13. Henry Obermeyer, *Stop that Smoke!* (New York and London, 1933), p. 189.

14. Hagley Library, Accession 1534 (Savery Family Papers), Box 7, Folder 11, Parsons Engineering Company, Notes on the Combustion of Coal and the Production of Smoke [undated], p. 6.

15. Ibid., p. 9.

16. Hagley Library, Accession 915 (William H. Savery Papers), Box 3, Folder 5, Report on Preliminary Tests of the Parsons Complete Combustion System to F. P. Fish, Esq. by Arthur D. Little, Boston, Mass., May 7th, 1908, p. 17.

17. Ibid., p. 14.

18. Ibid., p. 25.

19. Ibid., p. 14.

20. L. P. Breckenridge, *How to Burn Illinois Coal without Smoke* (University of Illinois Engineering Experiment Station, Bulletin Nr. 15) (Urbana, 1907), p. 15.

21. D. T. Randall and H. W. Weeks, *The Smokeless Combustion of Coal in Boiler Plants* (United States Geological Survey Bulletin 373) (Washington, 1909), p. 6.

22. A. L. Ayman, "Smoke Abatement," *Harper's Weekly* vol. 56 no. 2916 (9 November 1912), p. 23.

23. City of St. Louis, *Report of the Special Committee on Prevention of Smoke* (St. Louis, March 8th, 1892), p. 35.

24. Victor J. Azbe, "Smokeless Combustion in Domestic Heating Plants," *Mechanical Engineering* 51 (1929), pp. 761–764; p. 762.

25. R. R. Hillman, "Abating a Smoke Nuisance," *Power Plant Engineering* 26 (1922), pp. 463–465; p. 463.

26. William H. Bryan, "Smoke Abatement," *Cassier's Magazine* 19 (1900/01), pp. 17–24; p. 17; William H. Bryan, "The Problem of Smoke Abatement," *Scientific American Supplement* no. 1597 (August 11, 1906), pp. 25582–25584; p. 25582.

27. *Proceedings of the Engineers' Society of Western Pennsylvania* 8 (1892), p. 66.

28. Stadtarchiv Leipzig Hauptverwaltungsamt chap. 19, no. 11, vol. 1, p. 102. All translations from German by the author (Frank Uekoetter).

29. O. P. Hood, "Practical Suggestion on Smoke Abatement," *City Manager Magazine* vol. 8 no. 6 (June 1926), pp. 19–23.

30. Staatsarchiv Hamburg, 111-1 Cl. VII Lit. Q d Nr. 210 b vol. 1 b, document 1.

31. On the term *jurisdiction* and its implications, see Andrew Abbott, *The System of Professions. An Essay on the Division of Expert Labor* (Chicago and London, 1988).

32. Raymond R. Tucker Smoke Abatement Collection, Washington University Archives, St. Louis, Series 4 Box 1, Folder "Speeches — 1938", R. R. Tucker, Radio Talk — K.S.D., December 20, 1938, p.

33. Hagley Library, Pamphlet File, American Fuel Saving Company, Fuel Saved, Smoke Prevented (Philadelphia, n.d. [1899?]).

34. Cf. Frank Uekoetter, "Divergent Responses to Identical Problems: Businessmen and the Smoke Nuisance in Germany and the United States, 1880–1917," *Business History Review* 73 (1999), pp. 641–676.

35. Journal of the Franklin Institute 143 (1897), p. 414.

36. Gordon D. Rowe, "Experiences in Smoke Abatement Work," *Manual of Smoke and Boiler Ordinances and Requirements in the Interest of Smoke Regulation.* Published under the Auspices of the Public Service Committee of the Smoke Prevention Association, 1922 Edition (1922), pp. 55, 57.

37. H. B. Meller and L. B. Sisson, *The Municipal Smoke Problem. A Brief Treatise on the Products of Combustion of Fuels and their Effects* (Pittsburgh, 1935), p. 18.

38. Raymond R. Tucker, Smoke Abatement Collection, Washington University Archives, St. Louis, Series 4 Box 1, Folder "Speeches — 1939", Speech delivered before National Association of Power Engineers — Broadview Hotel, East St. Louis, Ill., September 9, 1939, p. 7.

39. Inspection of Boilers and Elevators and Smoke Abatement of the City of St. Louis, Annual Report for the Fiscal Year 1910–11, p. 8.

40. Department of Health, City of Chicago, Report for the Fiscal Year 1910–11, p. 8.

41. St. Louis Smoke Abatement Department, Smoke Abatement in Saint Louis. Report to the Mayor by the Smoke Abatement Department, March 1, 1909, p. 5.

42. Herbert M. Wilson, "The Cure for the Smoke Evil," *American City* 4 (1911), pp. 263–267; p. 265.

43. Bunte, "Die Massnahmen gegen Rauchbelästigung," *Hygienischer Führer durch die Haupt- und Residenzstadt Karlsruhe*. Festschrift zur XXII. Versammlung des deutschen Vereins für öffentliche Gesundheitspflege, hg. durch die Stadt und in deren Auftrag redigiert von R. Baumeister (Karlsruhe, 1897), pp. 102–103.

44. Chicago Department of Health, *Biennial Report of the Department of Health of the City of Chicago for the Years 1897 and 1898* (Chicago, undated), p. 135.

45. City of St. Louis, *Report of the Special Committee on Prevention of Smoke* (St. Louis, March 8, 1892), p. 23n.

46. Rochester Chamber of Commerce, Committee on Smoke Abatement, *The Abatement of Smoke* (1911), p. 10.

47. Chicago Department of Health, *Biennial Report of the Department of Health of the City of Chicago being for the Years 1895 and 1896* (Chicago, 1897), p. 306.

48. Chicago Historical Society, Illinois Manufacturers' Association, Box, 1 Folder 7, Glenn to Alling, May 15, 1902, p. 1

49. Charles A. L. Reed, *The Smoke Campaign in Cincinnati. Remarks before the National Association of Stationary Engineers, Cincinnati, July 10, 1906* (undated), p. 9.

50. Hauptstaatsarchiv Düsseldorf Regierung Köln Nr. 2213, Polizei-Verwaltung Mülheim to the Regierungspräsident of Cologne, October 28, 1902.

51. Badisches Generallandesarchiv Karlsruhe Abt. 233/4880, Ministerium des Innern to the Staatsministerium, July 28, 1892.

52. Geheimes Staatsarchiv Preußischer Kulturbesitz Berlin Rep. 120 BB II a 2 no. 28 Adh. 1 vol. 2 p. 255.

53. Staatsarchiv Ludwigsburg E 162 I no. 1419 doc. 208, 214, 222, 227; F 201 Bündel 680, file "betreffend den patentirten Ruß- und Funkenfänger des Fabrikanten B. Löffler in Frankfurt a/M."; Stadtarchiv Stuttgart Depot B, C XVIII 4 vol. 1 no. 1 Fasz. 162.

54. Stadtarchiv Nürnberg C 7/I Nr. 7161 vol. 1 p. 21r, 84, 92; Staatsarchiv Bremen 6,40 E 1 no. 17, Mitteilung des Senats 1888, p. 94n, dito 1889, p. 117, dito 1891, p. 217, dito 1892, p. 152; 4,14/1 IX.D.1.bc no. 16, Fabrikeninspektor Wegener to the Polizeidirektion, October 22, 1893; 3-G.4.a no. 49, Polizei-Akte betreffend Einführung der rauchverbrennenden Donneley-Dampfkesselfeuerung; 3-G.4.a no. 52, Senatskommission für Reichs- und auswärtige Angelegenheiten zu Bremen to the preußischer Minister der öffentlichen Arbeiten, October 27, 1893.

55. Archives of Industrial Society, University of Pittsburgh 83:7 Series III FF 19, Annual Report, Pittsburgh Bureau of Smoke Regulation, 1915, p. 4.

56. *Proceedings of the Eighth Annual Convention of the International Association for the Prevention of Smoke held at Pittsburgh, Pennsylvania, September 9–12, 1913* (Pittsburgh, undated), p. 75.

57. Obermeyer, *Stop that Smoke*, p. 189.

58. Joachim Raschke, *Soziale Bewegungen. Ein historisch-systematischer Grundriß*, (Frankfurt and New York, 1988) p. 376.

59. Cf. Uekoetter, *Divergent Responses*, and Uekoetter, *Von der Rauchplage*, for a more detailed discussion.

60. Paul P. Bird, *Report of the Department of Smoke Inspection, City of Chicago* (Chicago, 1911), p. 110n.

61. Thomas E. Donnelly, "Smoke Abatement in Chicago," *Papers read at the Smoke Abatement Conferences March 26, 27 & 28, 1912.* Edited by the Coal Smoke Abatement Society (London, undated), pp. 51–58; p. 56.

62. John O'Connor, Jr., "Civic Phases of the Smoke Problem," *National Municipal Review* 5 (1916), pp. 299–303; p. 302. On the origins of this article, see Archives of Industrial Society, University of Pittsburgh 83:7 Series I FF 19, Heath to O'Connor, July 21, 1915.

63. Hood, "Practical Suggestion," p. 22.

64. Missouri Historical Society, St. Louis, Citizens' Smoke Abatement League, Miscellaneous records, publicity material, pamphlets, etc., Report of Operations of Division of Smoke Regulation, Department of Public Safety, October 15, 1937 to October 15, 1938, p. 5.

65. Ayman, "Smoke Abatement," p. 23.

66. University of Illinois at Chicago, University Library, Special Collections, Woman's City Club of Chicago Folder 19, Bulletin vol. 16 no. 5 (January 1927), p. 185.

67. Cf. Uekoetter, "Divergent Responses," pp. 671–673. For more information on the association, see Frank Uekoetter, "Umweltschutz in den Händen der Industrie – eine Sackgasse? Die Geschichte des Hamburger Vereins für Feuerungsbetrieb und Rauchbekämpfung," *Zeitschrift für Unternehmensgeschichte* 47 (2002), pp. 198–216.

68. Staatsarchiv Hamburg 111-1 Cl. VII Lit. Q d Nr. 210 b vol. 2, Bericht des Vereins für Feuerungsbetrieb und Rauchbekämpfung in Hamburg über seine Tätigkeit im Jahre 1906, p. 8.

69. Ferdinand Haier, "Die Rauchfrage, die Beziehungen zwischen der Rauchentwicklung und der Ausnutzung der Brennstoffe, und die Mittel und Wege zur Rauchverminderung im Feuerungsbetrieb," *Zeitschrift des Vereines deutscher Ingenieure* 49 (1905), pp. 20–26, 83–88, 167–172; p. 21.

70. E. Nies, "Bericht des Vereins für Feuerungsbetrieb und Rauchbekämpfung in Hamburg über seine Tätigkeit im Jahre 1910," *Rauch und Staub* 1 (1910/11), pp. 275–276; p. 275.

71. Borgmann, "Erfolge in der Bekämpfung der Rauch- und Russplage an Dampfkesselfeuerungen in Linden, Hannover," *Zeitschrift für Heizung, Lüftung und Beleuchtung* 8 (1903/04), pp. 256–260; p. 258; Albano Brand, "Über rauchlose Feuerungen," *Sitzungsberichte des Vereins zur Beförderung des Gewerbfleißes* 1903, pp. 83–88; p. 88; Bernhard Rund, *Die Gefahren der Rauchplage und die Mittel zu ihrer Abwehr. (Ein Mahnwort zur Kohlenverschwendung)* (Wien, 1907), p. 9; C. Haage, "Das Rauchen der Fabrikkamine," *Mittheilungen aus der Praxis des Dampfkessel- und Dampfmaschinen-Betriebes* 6 (1883), pp. 126–127, 135–141; p. 135; Adolf Dosch, *Die Rauchplage und Brennstoffverschwendung und deren Verhütung* (Leipzig, 1920), p. 23.

72. Bird, *Report*, p. 129.

73. Ibid., p. 111.

74. *St. Louis Post-Dispatch*, July 12, 1911, p. 2, c. 4.

75. Charles H. Benjamin, "Smoke Prevention in the Power House," *Cassier's Magazine* 31 (1906/07), pp. 339–352; p. 342.

76. J. W. Henderson, "Smoke Abatement in Pittsburgh," *Domestic Engineering* 78 (1917), pp. 45–47; p. 45.

77. Wilson, *Cure*, p. 264.

78. Osborn Monnett, "Chicago Smoke Inspection Department Gets Action," *Industrial World* 48 (1914), pp. 131–132; p. 131.

79. *St. Louis Globe-Democrat*, January 15, 1916, p. 5, c. 2.

80. Matthew Nelson, "Smoke Abatement in Cincinnati," *American City* 2 (1910), pp. 8–10; p. 10.

81. Rochester Chamber of Commerce, Smoke Abatement Committee, *The Smoke Shroud. How to Banish It* (undated [1915?]).

82. Imogen B. Oakley, "Smoke is Waste Fuel," *The Civic Club Bulletin* [Philadelphia] 11 (January 1918), pp. 6–7; p. 6.

83. Stadtarchiv Aachen Cap. 15 no. 2 vol. 1 doc. 15, 30–33, 35, 37–42.

84. Stadtarchiv Leipzig chap. 35 no. 2 vol. 8 p. 248–249, 263, ibid. vol. 10 p. 1; *Mittheilungen aus der Praxis des Dampfkessel- und Dampfmaschinen-Betriebes* 17 (1894), p. 27; Staatsarchiv Hamburg 111-1 Cl. VII Lit. Q d Nr. 210 b vol. 1 b, doc. 42b, p. 42.

85. Staatsarchiv Bremen 3-G.4.a Nr. 334, Gewerbeaufsichtsamt Bremen to Senatskommission für das Gewerbeaufsichtsamt, 10 Nov. 1920; 4,14/1 IX.D.1.bc no. 16, Fabrikeninspektor Wegener to the Polizeidirektion, October 22, 1893.

86. Geheimes Staatsarchiv Preußischer Kulturbesitz Berlin Rep. 120 BB II a 2 no. 28 Adh. 2 vol. 2 p. 52–53.

87. Badisches Generallandesarchiv Karlsruhe Abt. 237/35385, Fabrikinspector to the Ministerium des Innern, May 7, 1886.

88. Geheimes Staatsarchiv Preußischer Kulturbesitz Berlin Rep. 120 BB II a 2 no. 28 Adh. 2 vol. 2 p. 133.

89. Ibid. vol. 5, Antrag für das Etatjahr 1906, Ministerium für Handel und Gewerbe to the Finanzminister, 30 Jul 1906; ibid. vol. 10, circular of the Ministeriums für Handel und Gewerbe, February 26, 1913.

90. Geheimes Staatsarchiv Preußischer Kulturbesitz Berlin Rep. 120 BB II a 2 no. 28 Adh. 2 vol. 3, Ministerium für Handel und Gewerbe to the Zentralverband der Preußischen Dampfkessel-Überwachungsvereine, November 17, 1902, Ministerium für Handel und Gewerbe to the Finanzministerium, August 26, 1903.

91. *Glückauf* 47 (1911), p. 1532n.

92. Geheimes Staatsarchiv Preußischer Kulturbesitz Berlin Rep. 120 BB II a 2 no. 28 Adh. 5 vol. 3, Ministerium für Handel und Gewerbe to the Zentralverband der preußischen Dampfkesselüberwachungsvereine, January 3, 1913.

93. Rebs, "Rauchverhütung," *Zeitschrift für Dampfkessel- und Maschinenbetrieb* 27 (1904), pp. 174–177, 186–188; p. 176.

94. Interestingly, a leaflet advertising a training course noted that the program was designed "to instill a sense of responsibility" among the fireman. (Stadtarchiv Frankfurt am Main, Magistratsakte T 726, p. 117.)

95. Badisches Generallandesarchiv Karlsruhe Abt. 237/35385, Badische Gesellschaft zur Ueberwachung von Dampfkesseln to the Ministerium des Innern, September 14, 1885 and January 25, 1894; Hauptstaatsarchiv Düsseldorf Regierung Düsseldorf Nr. 25055, Gewerbeinspektion Essen to the Regierungs- und Gewerberat zu Düsseldorf, April 21, 1899, p. 6n.

96. Staatsarchiv Hamburg 111-1 Cl. VII Lit. Q d Nr. 210 b vol. 6, Bericht des Vereins für Feuerungsbetrieb und Rauchbekämpfung in Hamburg über seine Tätigkeit im Jahre 1917, p. 5.

9

Fixing the Weather and Climate: Military and Civilian Schemes for Cloud Seeding and Climate Engineering

JAMES R. FLEMING

Cloud Wars

> The nation which first learns to plot the paths of air masses accurately and learns to control the time and place of precipitation will dominate the globe
>
> General George C. Kennedy,
> *Commander of the Strategic Air Command, 1947*

In 1947 meteorologists in the Joint Research and Development Board advised the U.S. Defense Department that weather control by cloud seeding was "entirely feasible" and that the perfection of this technique would have major tactical, strategic, and economic implications. If researchers were right, a small amount of nucleating agent could cause a chain reaction in clouds that would release as much energy as an atomic bomb. Such a weather weapon could be used surreptitiously and without radioactive fallout; moreover, it would be unidirectional in that clouds seeded over Europe would be carried by the westerly winds over the Soviet Union. Note the analogies to atomic weapons: cloud seeding nuclei, cause chain reactions forming a destructive mushroom-shaped cumulonimbus cloud with the energy of an atomic bomb; the fall-out — precipitation — is clean however (not radioactive). Such weapons might also be deployed surreptitiously.[1]

Military planners generated scenarios that included hindering the enemy's military campaigns by causing heavy rains or snows to fall along lines of troop movement and on vital airfields, using controlled precipitation as a delivery system for biological or radiological agents, diverting precipitation from one large geographic area to another to disrupt (or improve) the agricultural economy of the nation concerned, and altering the global climate for military purposes. Tactical possibilities included dissipating cloud decks to enable visual bombing attacks on targets, opening airfields closed by low clouds or fog, and relieving aircraft icing conditions. In military circles cloud seeding was seen as the trigger that could release the violence of the atmosphere against an enemy or tame the winds in the service of an all-weather air force. Based on the military and economic implications of the technique and the powers it promised its masters, meteorologists advised the military to launch an "intensive research and development effort."[2] For three decades, until 1977 when the United States signed a United Nations' Convention condemning the use of environmental modification as a weapon of war, the U.S. military enthusiastically supported research on new technologies of weather modification and control.

Such patronage should come as no surprise for indeed, modern meteorology has overwhelming ties to the military.[3] During World War II the U.S. Army Air Forces and the U. S. Navy trained approximately 8,000 weather officers. Personnel of the army's Air Weather Service (AWS), an agency nonexistent in 1937, numbered 19,000 in 1945. About 4,500 of this total were officers. Even after demobilization the AWS averaged approximately 11,000 soldiers during the Cold War and Vietnam eras.[4] In 1954 an NSF (National Science Foundation) survey of 5,273 professional meteorologists in America revealed that 43 percent of them were still in uniform on active duty, 25 percent held U. S. Air Force reserve commissions, and 12 percent were in the naval reserve. Thus almost a decade after the war, 80 percent of American meteorologists still had military ties.[5] Most of them were operational forecasters who had received their basic training in meteorology during the war. In 1958 the Committee on Meteorology of the National Academy of Sciences observed that, "the percentage of meteorologists with doctor's degrees was the lowest of any major scientific group, while the percentage of persons with no degree at all who were designated as meteorologists was the highest of any major scientific group."[6] In 1965, two decades after the war, the military had a larger budget than the Weather Bureau ($189 million compared to $104 million) and three times as many people (14,300 compared to 4,500) in weather service and research. The total meteorological budget of the U. S. Department of Defense (DoD) was $188.9 million in 1965. This was larger than that of any other government agency.

The General Electric (GE) Experiments and Project Cirrus

The modern era of weather modification began in the summer of 1946 at the General Electric Research Laboratory in Schenectady, New York, when

Vincent Schaefer dropped a block of dry ice into a home freezer unit and, to his surprise, instantly transformed a cold vapor cloud into millions of tiny ice crystals. After some rough calculations, Schaefer tossed six pounds of dry ice out the window of a rented plane and seeded a cold cloud over Greylock peak in the Berkshires, creating ice crystals and fall streaks of snow.[7] According to Schaefer's laboratory notebook, "It seemed as though [the cloud] almost exploded, the effect was so widespread and rapid ..."[8] Within a year Bernard Vonnegut (yes Kurt's brother) of the Massachusetts Institute of Technology (MIT), who had come to GE to count the crystals, discovered that silver iodide smoke also *seeded* supercooled clouds. Completing the cloud seeding triumvirate at GE was Nobel prize winner Irving Langmuir, senior scientist and an outspoken, enthusiastic promoter and popularizer of large-scale weather control.[9] In the press and before the meteorological community, Langmuir expounded his sensational vision of large-scale weather control: of the arid Southwest being changed into fertile farmland and of cloud seeding preventing "all ice storms, all storms of freezing rain, and icing conditions in clouds."[10]

An extremely optimistic announcement of progress in weather modification appeared in the *General Electric Annual Report* for 1947: "Further experiments in weather control led to a new knowledge which, it is believed now, will result in *inestimable* benefits for mankind."[11] But cloud seeding was becoming a controversial issue and Langmuir's exaggerated claims threatened to take the company into litigious territory, far beyond the limits of normal corporate support for research.[12] When one of Schaefer's cloud seeding attempts coincided with an eight-inch snowfall in upstate New York (earlier the Weather Bureau had forecast "fair and warmer"), Langmuir was quick to claim that cloud seeding had triggered the storm. He further claimed that chain reactions could be set off in warm cumulus clouds, that in one field trial a hurricane had changed direction within six hours because of seeding, and that in general all meteorologists needed to do was find the proper trigger to release the immense amounts of energy stored in the atmosphere.[13]

GE lawyers, fearing a deluge of property damage and personal inconvenience suits, tried to silence Langmuir.[14] C. Guy Suits, the director of research at GE, hurriedly halted outdoor experimentation on cloud seeding. He instructed Langmuir's team to work with Project Cirrus, a classified project conducted by the U. S. Army Signal Corps, the Office of Naval Research and the U. S. Air Force over a 1,000-square-mile area north of Schenectady. As stated in the GE contract, the general purposes of the project were (a) "to search for fundamental knowledge and a better understanding of the physics and chemistry of hydrometeor formation as an aid in operational forecasting of both local and long range weather conditions" and (b) "to investigate the practicability of and the means for cloud modification for military purposes."[15] The *Harvard Law School Record* reported, "Today 'Project Cirrus' has an annual budget of $750,000 from military and naval funds because of its war implications — bogging down enemy troops in snow and rain, clearing airfields of fog at lowest cost, and infecting induced storms with bacteriological and radiological materials. The Battle of the Bulge, in which the Nazis mobilized and attacked under supercooled fog, could have been much altered by a few pounds of dry ice."[16]

Langmuir and Schaefer served only in advisory roles in Project Cirrus. GE's contract stipulated that, "the entire flight program shall be conducted by the government, using exclusively government personnel and equipment, and shall be under the exclusive control of such government personnel." Suits notified his staff, "that it is essential that all of the G.E. employees who are working on the project refrain from asserting any control or direction over the flight program. The G.E. research laboratory responsibility is confined *strictly* to laboratory work and reports."[17] GE's corporate posture was that the whole matter properly belonged to the government, and that the government, by suitable legislation, should both regulate the inducing of rainfall and indemnify for loss any contractor acting on the government's behalf — especially themselves. Secretary of Defense James Forrestal asked Congress for a law "to protect contractors engaged in cloud modification experiments against claims for damages by third parties."[18] No legislation to this effect, however, was forthcoming.

Under Project Cirrus, between 1947 and 1952, scientists carried out about 180 field experiments involving modification of cold stratus clouds, warm and cold cumulus clouds, and a tropical storm. Although mountains of photographic and other data were collected and analyzed, the response of the atmosphere to seeding was erratic and no definitive measure of the efficacy of artificial nucleating agents was obtained.[19] The results from several experimental runs were spectacular, however, and the DoD decided to continue funding projects, which continued and expanded the work of Project Cirrus.

One such project, the joint Air Force-Weather Bureau Cloud Physics Research Project, designed to study the synoptic aspects of cloud physics and test the results of Project Cirrus, found that seeding did indeed produce striking visual changes in clouds, including dissipation of cold stratus decks. However, experiments with clouds over Ohio in 1948 and over California and the Gulf states in 1949 led the researchers to conclude that cloud seeding could not initiate self-propagating storms or relieve drought. The Weather Bureau spent $85,000 on the project in 1948 and $100,000 in 1949. The U. S. Air Force supplied aircraft, personnel, and ground radar facilities.[20]

Experiments in New Mexico and Secret Military Projects

Langmuir's most fantastic claim was that changes in the weather across the continent had been caused by a single silver iodide generator in New

Mexico in 1950.[21] The experiment, which coincided with severe flooding in the Ohio Valley and resulted in widespread property damage and fatalities, came to the attention of the highest echelons in the government and resulted in a number of secret projects. Noted meteorologist Sverre Petterssen of the University of Chicago explained the situation as follows:

For no profound reason [Langmuir] had left a silver-iodide generator somewhere in New Mexico and made arrangements with a local person to "burn" the generator on a weekly schedule. Using a set of readily available weather reports, Langmuir found that the rainfall had begun to vary in a weekly rhythm. The amazing thing was that the response was not just local; it was nationwide and might well be of hemispheric proportions. Langmuir, and many with him, concluded that the weekly injection of silver iodide from a single generator in New Mexico had excited a hitherto undiscovered natural rhythm of the atmosphere ...[22]

Langmuir had forgotten, or perhaps was unaware, that the atmosphere frequently exhibits a *natural* seven-day periodicity.

Nevertheless, the technology seemed of such great potential, especially to military aviation, that Vannevar Bush, former head of the Office of Scientific Research and Development (OSRD) in World War II, brought the issue to the attention of Secretary of Defense George C. Marshall and General Omar Bradley, chairman of the Joint Chiefs of Staff. Bradley immediately convened a "cushion committee" to serve as a buffer between the defense establishment and the scientific community. This committee, consisting of an admiral, a general, and the chief of the U.S. Weather Bureau, Francis F. W. Reichelderfer, in turn appointed a special scientific committee chaired by Sverre Petterssen, director of scientific weather services for the U.S. Air Force, called the "Ad hoc Committee on Artificial Cloud Nucleation" — a name, according to Petterssen, "that did not suggest interest in secret weapons. To add camouflage, Dr. Alan T. Waterman, Director of the National Science Foundation, was appointed a member."[23]

At the direction of the cushion committee, and with the secret hope that a secret weapon might emerge from this technology, Petterssen's ad hoc group conducted a brief survey of the state of the field and recommended a program of five statistically controlled experiments to begin in 1952 "to clarify major uncertainties."

1. The seeding of extratropical cyclones, designated Project Scud, was sponsored by the Office of Naval Research, with flight operations performed by the Navy Hurricane Reconnaissance Squadron, and data analysis under the direction of Dr. Jerome Spar of New York University. The project sought to evaluate Langmuir's claim that large-scale weather systems could be controlled, but the experimenters concluded that, "the seeding in this experiment failed to produce any effects which were large enough to be detected against the background of natural meteorological variance."[24]

2. Modification of convective clouds was sponsored by the Air Force Cambridge Research Center, under contract with the University of Chicago. The Air Force Research and Development Command supplied flight services. Dr. Horace R. Byers, head of the well-developed and essentially permanent cloud physics laboratory at the Chicago Midway Laboratory of the University of Chicago, led the research team.[25] Because of ongoing funding from DoD, his physics lab was one of the better-equipped and more completely staffed facilities in the world. The stated purpose of this investigation was to determine to what extent cold and warm cumulus and cumulonimbus clouds could be modified by artificial nucleation. To obtain as many cumulus clouds as possible for testing, the group worked in the Caribbean in the winter and in the Midwest in the summer. Caribbean clouds were seeded with water from a 450-gallon tank in the bomb bay of an air force B-17. Statistical analysis indicated that many more seeded clouds had radar echoes than unseeded ones, and the probability of occurrence of an echo in a cloud was approximately doubled by seeding. Both water and dry ice were used as seeding agents in Illinois and Missouri during the summers of 1953 and 1954, but the numbers of clouds seeded were deemed too small to yield significant results.
3. The U.S. Air Force, under contract with Stanford Research Institute, also examined the physics of ice fogs and their relation to weather conditions at air bases in Alaska. The study showed that most ice fogs developed from local sources of water and pollution, such as chimneys, smoke stacks, power plants, motor vehicles, and aircraft during warming-up operations.
4. Experiments in the dissipation of cold stratus and fog, sponsored by the Army Signal Corps Engineering laboratories, in large part substantiated the results of Project Cirrus.
5. The Army also hired Arthur D. Little, Inc. to explore techniques for modifying warm stratus and fog. Under Vonnegut's leadership, attempts were made to cause precipitation of suspended cloud droplets by generating electrostatic forces or by adding chemical agents to the fogs to increase their drop sizes. Vonnegut worked for Little from 1952 to 1967 on meteorological and other problems. Apparently, controversies surrounding weather modification were so great that the company, like GE, did not emphasize this line of research.[26]

Although this series of experiments had better statistical controls than did Project Cirrus, they did not receive much attention and they most definitely did not lead to new military applications. The experiments ended in 1954, but the final report was not published until 1957 when it appeared in a limited circulation monograph of the American Meteorological Society. All of the reports needed to pass a security review; most of the delay, however, was caused by the Petterssen committee itself, which collected all five reports and insisted on revisions before publishing any of them. The reports were out of date as they left the press.[27] Moreover, the meteorological community discounted the experiments because of their brevity, their inconclusive results, and the primitive state of the art of instrumentation. It seemed that the design of the experiments was not sufficiently sophisticated to filter out the natural variability of the atmosphere.[28] The optimistic position on the topic adopted by the Council of the American Meteorological Society in 1950 was no longer supported by the meteorological community.[29] According to noted meteorologist J. Murray Mitchell, the negative results of the Artificial Cloud Nucleation Committee, "encouraged the meteorological profession to 'go underground' in weather control research," and concentrate "on fundamental studies of cloud and precipitation physics."[30]

The Controversies Continue

While the military and Weather Bureau projects were struggling for results, a determined and enthusiastic band of private meteorological entrepreneurs, operating primarily in the West and Midwest, succeeded in placing nearly 10 percent of the land area of the country under commercial cloud seeding at an annual cost to farmers and municipal water districts of $3 to $5 million.[31] The spread of this technique generated numerous public controversies that pitted Langmuir, the entrepreneurs, and their clients against Weather Bureau skeptics and parties claiming damages purportedly caused by cloud seeding. For example, in 1951 New York City was facing 169 claims totaling over $2 million from Catskill communities and citizens for flooding and other damages attributed to the activities of a private rainmaker, Wallace Howell. The city had hired Howell to fill its reservoirs with rain, and at least initially claimed that Howell had succeeded. When faced with the law suits however, city officials reversed their position and commissioned a survey to show that the seeding was ineffective. Although the plaintiffs were not awarded damages, they did win a permanent injunction against New York City that ceased further cloud seeding activities; further litigation stopped just short of the Supreme Court.[32]

During the western drought in the early 1950s, Irving Krick,[33] private weather consultant and promoter of a controversial system of ultra–long-range forecasting, began cloud seeding operations for large agricultural concerns. His clients included wheat farmers, ranchers, and stream-flow enhancement projects on the Salt River in Arizona and the Columbia

River in the Pacific Northwest. In the later project, Krick was credited by the Bureau of Reclamation with an 83 percent enhancement of the river flow, while the Weather Bureau considered this claim meaningless and sought to discredit him whenever possible. At the height of his operations, Krick's company was conducting seeding operations covering 130 million acres of western lands.

Of greatest concern to the military was the possibility that such unregulated civilian cloud seeding could contaminate the atmosphere and render military tests inconclusive or worse, that it could generate such controversy that all testing could be stopped by litigation. In the opinion of military legal counsel, it was "considered essential to protect the research program from stoppage by injunction or other pressure."[34] F. W. Reichelderfer, head of the U.S. Weather Bureau, advised the DoD that control of weather could be even more controversial than control of atomic energy and might offend some religious groups; unbridled controversy could threaten "further development and progress in this field." Vannevar Bush thought civilian requests for participation in application of cloud and weather modification should be referred to the Weather Bureau as the coordinating agency.[35] Suits suggested the creation of a central governmental agency to control weather modification. A Harvard study group agreed and advocated spending a portion of the Atomic Energy Commission's budget on weather modification research.[36] A bill (S. 222) introduced in the 81st Congress proposed a "Weather Control Commission," to be established and organized substantially along the lines of the Atomic Energy Commission, with advisory and liaison committees. The DoD, seeing this as a threat to its autonomy, was strongly opposed to any laws creating such a new agency or centralizing authority in an existing agency.[37]

The debate over private and public experiments in weather control prompted Congress to establish an Advisory Committee on Weather Control in 1953. It was chaired by a presidential appointee, retired U.S. Navy Captain Howard T. Orville.[38] The committee's report, issued in 1958, was cautiously optimistic, concluding that increases of 10 to 15 percent in rainfall were induced by seeding spring and winter storms in the mountainous areas of the western United States. The committee recommended more vigorous government support for basic meteorological research, specifically in solar influences, global air circulation, dynamics of cloud motion, and the origin and movement of large-scale storms. It also recommended a dramatic increase in manpower. In the light of these recommendations, NSF became, in 1958, the lead agency for federally supported research on weather modification.[39]

The Cold War Era: Strategy and Tactics

Although the time scales are different by many orders of magnitude, the total amount of energy released by a single thunderstorm is equal to that of a twenty-kiloton atomic bomb. Moreover, a mature hurricane of moderate strength and size releases as much energy in a day as that of about 400 twenty-megaton hydrogen bombs. Such impressive numbers, despite the technical uncertainties involved in large-scale weather modification, made comparisons between weather modification and nuclear weapons very popular.[40] Since weather warfare could perhaps be conducted cheaply, surreptitiously, and without polluting the environment with fallout, and since the Russians were probably going to master the technology anyway, it made good sense to military planners to attempt to control it (see Figure 9.1).

The classic Cold War pronouncement on weather prediction and control belongs to General George C. Kennedy, commander of the Strategic Air Command: "The nation which first learns to plot the paths of air masses accurately and learns to control the time and place of precipitation will dominate the globe."[41] Kennedy was not alone. The distinguished aviator-engineer Rear Admiral Luis De Florez, who developed synthetic training devices for navy fliers during World War II, had a similar opinion: "With control of the weather the operations and economy of an enemy could be disrupted ... [Such control] in a cold war would provide a powerful and subtle weapon to injure agricultural production, hinder commerce and slow down industry." De Florez advocated that government, "start now to make control of weather equal in scope to the Manhattan District Project which produced the first A-bomb."[42]

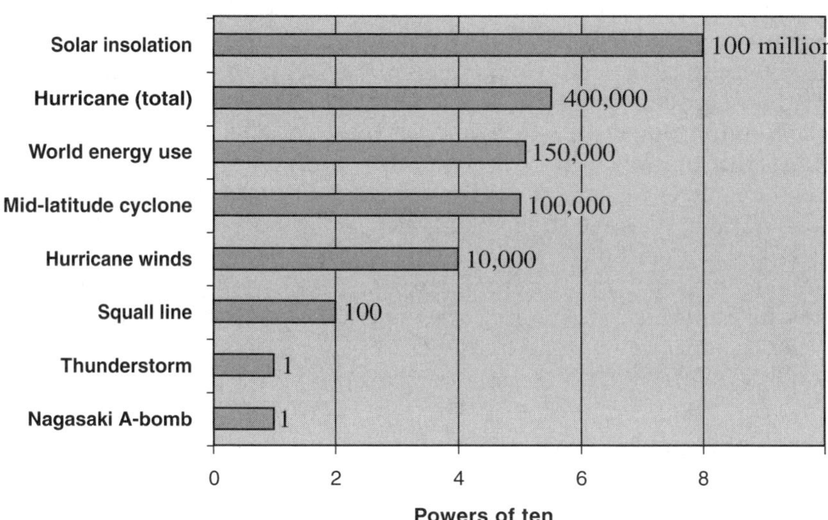

Fig. 9.1 Energy of Selected Phenomena Expressed in A-bomb Equivalents.

Pursuing this theme, Orville, Eisenhower's weather advisor, published an influential article in *Collier's* in 1954 that included scenarios for meteorological warfare. Planes would drop hundreds of balloons containing seeding crystals into the jet stream. Downstream, when fuses on the balloons exploded, the crystals would fall into the clouds, initiating rain and miring enemy operations. The Army Ordnance Department was investigating another technique: loading silver iodide and carbon dioxide into fifty-caliber tracer bullets that pilots could fire into clouds.[43] A more insidious technique would strike at the enemy's food supply by seeding clouds to rob them of moisture before they reached enemy agricultural areas:

> It is … conceivable that we could use weather as a weapon of warfare, creating storms or dissipating them as the tactical situation demands. We might deluge an enemy with rain to hamper a military movement or strike at his food supplies by withholding needed rain from his crops. … But before we can hope to achieve all the *benefits* I have outlined, hundreds of meteorological unknowns must be solved at a cost of possibly billions of dollars.[44]

Although in Orville's assessment, total weather mastery would be possible only after four decades of intensive research, the spin-offs from this work, when combined with the maturation of electronic computers, would provide a *completely accurate* system of weather forecasting, perhaps within a decade: "I think it entirely probable that, in 10 years, your daily weather forecast will read something like this: 'Freezing rain, starting at 10:46 A.M., ending at 2:32 P.M.' or 'Heavy snowfall, seven inches, starting today at 1:43 A.M., continuing throughout day until 7:37 P.M.'"[45] This sort of accuracy of prediction, even without weather control, would have major consequences for military operations. Although speculative and wildly optimistic, ruminations such as these from an official source and threats that the Soviets were aggressively pursuing weather control helped fuel the rapid expansion of meteorological research in all areas during the Cold War era. With the support of Waterman at NSF, the atmospheric sciences began a period of rapid growth. This is especially noticeable in the founding of many new university departments of atmospheric science and the University Corporation for Atmospheric Research, and in the building of the National Center for Atmospheric Research, a national laboratory in Boulder, Colorado.[46]

A Geophysical Arms Race

In January 1958, while Americans were still reeling from the psychological impact of the launch of the Soviet's first earth satellite, *Newsweek* informed its readers that there was a "new race with the reds" in the form of weather warfare. In the article, Orville indicated the need to keep ahead of the Russians was clearer than ever: "If an unfriendly nation gets into a position to control the large-scale weather patterns before we can, the result could even be more disastrous than nuclear warfare."[47] The article also quoted hot heads such as Edward Teller, an expert on hydrogen bombs but not on weather control, who told the Senate Preparedness Subcommittee: "Please imagine a world ... where [the Soviets] can change the rainfall over Russia ... and influence the rainfall in our country in an adverse manner. They will say, 'we are sorry if we hurt you. We are merely trying to do what we need to do in order to let our people live.'"[48] Cooler heads, such as Professor Henry G. Houghton, chairman of the Department of Meteorology at MIT, expressed the same concerns: "I shudder to think of the consequences of a prior Russian discovery of a feasible method of weather control ... An unfavorable modification of our climate in the guise of a peaceful effort to improve Russia's climate could seriously weaken our economy and our ability to resist."[49]

According to the Soviet academician Ye. K. Fedorov, the Soviet Union was engaged in a "struggle for meteorological mastery" over nature and, by analogy to the space race, over the United States:

> mankind is ever more rapidly approaching that stage in its interaction with nature in which it will require practically all the natural resources of the earth and will need the capability to cope with elemental phenomena on a large scale over the entire globe. In other words, man is becoming master of the earth. It is evidently no accident that our entry into space coincides with this development. There is hardly any need to argue that under these conditions all mankind must present a united front to the world around it. There is no other alternative. There is nothing absurd in dreams of a 'struggle for meteorological mastery.'[50]

Such mastery was but one aspect of the Soviet's "Great Plan for the Transformation of Nature," announced by Stalin in 1948. Under this plan the Soviet State Planning Commission controlled all activities devoted to expanding the Soviet economy by harnessing nature: opening new farmland, building new cities and hydroelectric installations, establishing shelter belts, and attempting to control the weather and climate. Science was seen as a useful ally in what was considered a "war against nature on a farflung front"; in such a struggle, the ability to control the weather and climate was considered particularly useful.[51]

By some estimates, the Soviet Union had three or four times more meteorologists than the United States. As noted earlier, in 1965 the U.S. military employed 14,300 meteorologists and the Weather Bureau, 4,500. In 1969 the president of the American Meteorological Society was informed by official Soviet representatives that the total number of employees in the Soviet Hydrometeorological Service was in excess of 70,000. An estimate published in 1974 placed the number of employees of the U.S. National Weather Service at 6,000 and the number of Soviet meteorologists at 75,000.[52]

The issue of a geophysical arms race did not go away. In the late 1960s Gordon MacDonald, professor of geophysics at the University of Southern California, Los Angeles (UCLA), Pentagon confidant, and member of President Johnson's Science Advisory Committee, saw weather control, even of severe storms such as hurricanes and typhoons, as just the beginning step in an escalating game of environmental and geophysical warfare. Potentially, belligerents could, for example, cut a hole in the ozone layer over a target area to let in lethal doses of ultraviolet radiation, manipulate the Arctic ice sheet to cause climatic changes or massive tidal waves, trigger earthquakes from a distance, and in general manipulate the planetary environment and its geophysics on a strategic scale.[53]

Nor was the discussion total science fantasy. Military planners, perhaps oversimplifying the problems involved, were interested in improved weather observations, understanding, prediction, and control to support both current and future operational capabilities. Weather and climate modification research included both computer modeling of the atmosphere and new weather satellites. Nile Blue, a computer model developed by Advanced Research Projects Agency (ARPA), was used to test the sensitivity of the climate to major perturbations, including Soviet tinkering and those caused by a major environmental war. *Newsweek* reported that the United States would soon have a series of "baby moons" of its own, and that one of the "eye-in-the-sky" satellites, designed by the Army Signal Corps, would be used to monitor the Earth's weather from space. Military weather satellites would be useful in routine operational forecasting, as support for the climate modelers, and for more esoteric missions such as monitoring changes in global weather patterns and heat budgets, and tracing the effects of nuclear tests. But while most Americans worried about the race with the reds and the military speculated about the use of climatic and geophysical modification techniques as strategic weapons in future wars, operational cloud seeding was being used tactically and covertly in a real war — over the jungles of Vietnam.[54]

Cloud Seeding in Indochina

In March 1971, nationally syndicated columnist Jack Anderson broke a story about air force rainmakers in Southeast Asia.[55] Several months later a document published in the *Pentagon Papers* confirmed that the government, after field trials in Laos, had begun cloud seeding operations in Vietnam in 1967 to reduce trafficability along enemy infiltration routes. This was done with the full and enthusiastic support of President Johnson. The code name of the operation was POPEYE.[56]

Between 1967 and 1972, during the rainy southwest monsoon season, Operation POPEYE attempted to create rain over portions of the Ho Chi

Minh Trail in North Vietnam, Laos, Cambodia, and South Vietnam.[57] The AWS conducted the mission out of Udorn Air Base, Thailand. Employing C-130 and other aircraft equipped with racks of up to 104 silver or lead iodide flares, the AWS flew over 2,600 sorties and expended almost 50,000 flares over a period of approximately five years at an annual cost of approximately $3.6 million. This information, from a top secret Defense Department briefing in 1974, was made public several days later by Senator Claiborne Pell.

Although by some reports, Operation POPEYE induced from one to seven inches of additional rain along the Ho Chi Minh Trail, no scientific data were collected to verify the claim.[58] General William Westmoreland thought there was "no appreciable increase" in rain from the project.[59] Even if the cloud seeding had produced a tactical victory or two in Vietnam (it did not), the extreme secrecy surrounding the operation and the subsequent denials and stonewalling of Congress resulted in a major strategic defeat for military weather modification. To illustrate the point, Westmoreland was one of only four general officers in Southeast Asia privy to the details of Operation POPEYE. Even the squadron commander at Udorn whose planes were being used was not informed of the mission.[60] The governments of Thailand, Laos, and South Vietnam were not informed, nor were the American ambassadors to those countries. Typical of the cover-up during this period was the AWS annual survey report on weather modification for 1971, which documented five fog dispersal projects in the United States and Germany, and a rainmaking effort in Texas, but contained no mention of Operation POPEYE.[61] The prime example of stonewalling came from Secretary of Defense Melvin Laird who told the Senate Foreign Relations Committee in 1972 that there was no cloud seeding going on over *North Vietnam*, but never mentioned that Operation POPEYE was still functioning over Laos, Cambodia and South Vietnam.[62]

Operation POPEYE, made public as it was at the end of the Nixon era, was called the Watergate of weather warfare.[63] But it was neither the first use of weather modification as a weapon of war, nor the first in Southeast Asia. On April 22, 1954 the French High Command announced, in connection with the besieged French forces at Dien Bien Phu that, "it will try to wash out Vietminh communication routes from Red China with man-made rainstorms as soon as cloud conditions permit."[64] Moreover, according to the testimony of a former CIA agent, the agency seeded clouds in South Vietnam as early as 1963 in an attempt to disperse demonstrating Buddhist monks when they noticed that the monks resisted tear gas but disbanded when it rained.[65] Some former DoD consultants were politicized by the revelations. One alleged (but Pentagon sources denied) that weather modification was also attempted against Cuba during 1969 and 1970 in an attempt to cause a damaging drought.[66] Another observed that the lesson of the Vietnam experience was not that rainmaking is an inefficient means for slowing logistical movement in jungle trails, but "that one can conduct covert operations using a new technology in a democracy without the knowledge of the people."[67] The dominant opinion, however, was that seeding clouds, like using Agent Orange or the Rome Plow, setting fire to the jungles or bombing the dikes over North Vietnam, was but one of many sordid techniques used in Vietnam.[68]

Some argued that environmental weapons were more humane than nuclear weapons. Others suggested that inducing rainfall to reduce trafficability was preferable to dropping napalm; as one wag put it, "make mud, not war." Philip Handler, president of the National Academy represented the mainstream of scientific opinion, however, when he observed: "It is grotesquely immoral that scientific understanding and technological capabilities developed for human welfare to protect the public health, enhance agricultural productivity, and minimize the natural violence of large storms should be so distorted as to become weapons of war."[69] Nevertheless, for every prophetic call to beat "swords into plowshares" (Isaiah 2:4), there were many more who advocated that plowshares should be beaten into swords (Joel 3:10).

The weather modification budget as reported by DoD averaged about $1.5 million per year in constant 1967 dollars and reached zero by 1979. In 1962, however, the only year ARPA reported its budget, the expenditures reported were an additional $1.8 million.[70] A request for the weather modification and cloud physics budgets of ARPA and its successor Defense Advanced Research Projects Agency (DARPA) from 1961 to 1978 filed under the Freedom of Information Act resulted in a "no record" response.[71] One unofficial source estimated that ARPA's budget averaged $3.3 million per year for weather modification research during the entire period.[72] The cost of operational rainmaking in Vietnam, hidden from official reports but included here, averaged another $3.6 million per year from 1967 to 1972, or more than double the reported military research budget. The cost of operating and maintaining the aircraft (estimated at $1 to 1.5 million per year) and salaries and support of military personnel engaged in cloud seeding are not included in these figures.

Outlawing Environmental Modification as a Weapon of War, 1972 to 1977

In 1972 Senator Claiborne Pell, prompted in large part by unofficial reports that the United States was modifying weather conditions in Southeast Asia as a part of its military operations, introduced a resolution calling upon the United States government to negotiate a convention prohibiting the use of environmental or geophysical modification activities as weapons of war.[73] Testifying at the Senate hearings, Richard J. Reed, president of the American Meteorological Society, cited earlier bans on chemical and biological warfare and atmospheric nuclear testing, and urged the government," to present for adoption by the United Nations General Assembly a resolution pledging all nations to refrain from using weather modification for hostile purposes." Citing a 1972 public policy statement of the society, he referred to the "present primitive state of knowledge" in the field and the difficulties of controlled experimentation during military operations. The testimony of other prominent atmospheric scientists stressed the need to protect open and peaceful international scientific cooperation.[74] Despite the opposition of the Nixon administration, the Senate adopted the resolution within a year by a vote of 82 to 10.[75] Representative Donald Fraser led a parallel effort in the House.

Feeling the growing pressure from Congress and the press, President Nixon directed a committee of the National Security Council (NSC) to undertake a study of possible international restraints on environmental warfare. In May 1974 the committee presented the following options to the president: (1) no restraints; (2) restraints on military use of environmental modification techniques having long-term, widespread or especially severe effects; (3) a comprehensive prohibition of hostile use.[76] The administration favored option two.

At the Moscow summit on July 3, 1974, just after Senator Pell had placed the top secret DoD briefing on cloud seeding in Vietnam in the public record, President Nixon and Soviet General Secretary Brezhnev signed a "Joint Statement Concerning Future Discussion on the Dangers of Environmental Warfare" expressing their desire to limit the potential danger to mankind from the use of environmental modification techniques for military purposes whose effects would be "widespread, long-lasting and severe."[77] This was essentially option number two of the NSC. Military restraint was limited to conjectural and highly impractical techniques of climatic and environmental modification having long-term, widespread or especially severe effects. By that language, more or less operational techniques of weather modification such as rainmaking and fog dispersal, whose effects were considered limited in time, place and effect, were excluded from the discussion; their use in warfare was implicitly legitimized.[78]

Within a month the Soviet Union, realizing the weakness of the U.S. position on cloud seeding in Vietnam and taking full advantage of the Watergate crisis, seized the diplomatic initiative by unilaterally bringing the issue of weather modification as a weapon of war to the attention of the United Nations. The Soviet proposal did not limit the treaty to a bilateral agreement, nor did it limit it to effects that were "widespread, long-lasting and severe." According to the Soviets, "It is urgently necessary to draw up and conclude an international convention to outlaw action to influence the environment for military purposes."[79] The draft convention unveiled by the Soviet Union in September of 1974 forbid contracting parties from using:

> meteorological, geophysical or any other scientific or technological
> means of influencing the environment, including weather and cli-
> mate, for military and other purposes incompatible with the main-
> tenance of international security, human well-being and health,
> and, furthermore, never under any circumstances to resort to such
> means of influencing the environment and climate or to carry out
> preparation for their use.[80]

The General Assembly, taking note of the Soviet draft convention, decided that the subject deserved further attention and, with the United States abstaining, voted to turn it over to the Conference of the Committee on Disarmament.[81] To avoid further embarrassment, President Ford insisted on the language of NSC's option number two. The final treaty, the U.N. Convention on the Prohibition of Military or any other Hostile Use of Environmental Modification Techniques, applied only to environmental modification. The qualifiers "widespread, long-lasting or severe" had found their way back into the convention.

The convention was opened for signature on May 18, 1977 in Geneva. It was signed initially by thirty-four states, including the United States and the Soviet Union, but did not enter into force until October 5, 1978, ironically, when the Lao People's Democratic Republic, where the American military had used weather modification technology in war only six years earlier, became the twentieth nation to ratify it.[82]

As the United States had hoped, the convention was vague and unenforceable. The AWS was of the opinion that the treaty's language was so vague that it did not affect its capabilities in weather modification at all, and Military Airlift Command was instructed to retain its capabilities in this area. For the military, the deciding factor was not the U.N. convention, but the fact that weather modification technology had "little utility" or "military payoff" as a weapon of war.[83] By 1978, the DoD claimed that its operational programs were directed solely at fog and cloud dispersal, while research funding continued in cloud physics, computer modeling, and new observational systems.

Yet Operation POPEYE was the mouse that roared, causing international embarrassment as the United States fought to tone down the wording of the convention. The close links between the meteorological community and the military, forged during World War II and the early Cold War, eroded over time. Faced with the complexities of the atmosphere, meteorologists did not develop reliable technologies of weather modification and control. Moreover, a new generation of university-trained meteorologists feared that the atmosphere of secrecy surrounding military projects would ultimately poison the atmosphere of international cooperation needed in meteorological research. Although military patronage since World War II had provided consistent support for several generations of meteorologists, approximately two-thirds had gone to basic weather analysis and forecasting, while only one-third had supported the various research specialties.

By the 1970s the atmospheric science community, no longer dominated by the veterans of World War II, had trained a new generation of civilian Ph.D.s who increasingly were funded by NSF and NASA. Their primary interest was in the advancement of the discipline, not in national and international political controversies. Many of them were aghast that the unproven weather modification technologies were garnering adverse publicity and generating such a high-level debate. They found it all quite threatening to the free international exchange of meteorological data, to ongoing international research programs such as GARP (the Global Atmospheric Research Programme), and to the future of peaceful international scientific cooperation.[84]

The Current Situation

Climate modification proposals have returned in force recently due to growing apprehension that global warming is real and will have real consequences. As the logic goes, if everyone is unwilling or unable to limit their greenhouse gas emissions, perhaps a few smart people can provide a technological fix for the climate.

In 1992 the Intergovernmental Panel on Climate Change concluded that "the unequivocal detection of the enhanced greenhouse effect from observations is not likely for a decade or more," a widely cited conclusion of the 1995 report was that "The balance of evidence suggests a discernible human influence on global climate."[85] On the social end, the United Nations Environmental Programme recently asked, "Are we overlooking the social and political implications of climate change?" pointing out that if scientific predictions about climate change hold true, it seems likely that political structures and social bonds will be subjected to additional stresses.[86] Several strategies of climate intervention are being pursued. The Montreal Protocol and the United Nations Framework Convention on Climate Change represent geopolitical interventions in the climate system. Many more policy initiatives are underway. Economics has also begun to play a role as taxes and incentives are put in place to reduce unwanted emissions. Meanwhile, green social engineers are attempting to convince the general public to save the planet by reducing, reusing, and recycling.

The most immodest of the new intervention strategies involves *geoengineering* massive technical fixes for the climate system. A 1991 National Academy of Sciences report, *Policy Implications of Greenhouse Warming*, advised that the United States should conduct research in schemes to cool the Earth if global warming gets out of hand. Proposals included orbiting a fleet of space mirrors or spraying sulfur dioxide into the stratosphere to reflect solar radiation back into space, turning the oceans into soupy green algae blooms to sequester excess carbon, or setting up gigantic "soot generators" to shade the Earth. Other scholars have taken a recent serious look at geoengineering and find it attractive because in their words, "Doubt about the prospects for cooperative abatement of global greenhouse gas emissions is a pragmatic reason to consider geoengineering, whose implementation requires fewer cooperating actors than abatement."[87] If this does not invoke apprehension, I don't know what will. As Jerome Namias pointed out in 1989, "the greenhouse effect is now firmly part of our collective angst, along with nuclear winter, asteroid collisions, and other widely bruited global nightmares."[88]

Lest we forget, agricultural, water conservation, and hydropower interests are conducting routine cloud seeding operations over about one-third of the area of the American West (see Figure 9.2).

On a more speculative level, three announcements were recently in the news: Beijing's Study Institute of Artificial Influence on the Weather has announced its intention of manipulating the weather to ensure optimum conditions for the 2008 Olympics; a private weather company in Florida has announced a new powder called Dyno-Gel that has the power to "suck the moisture out of a thunderstorm or weaken a hurricane"; and finally, a recent study by the U.S. Air Force claims that "in 2025, US aerospace forces can 'own the weather' by capitalizing on emerging technologies and focusing development of those technologies to war-fighting applications." In addition to traditional cloud seeding methods, the U.S. Air Force visionaries propose computer hacking to disrupt an enemy's weather monitors and models, and using nanotechnology to create clouds of microscopic computer particles that could block an enemy's optical sensors or guide smart weapons to their targets; the cost developing these clouds is to be borne by the private sector. In a recurring theme, the military points out that weather modification, unlike other approaches, "makes what are otherwise the results of deliberate actions appear to be the consequences of natural weather phenomena."[89]

Epilogue

As an episode in the history of science in the Cold War era, the story of weather control contains tragic, comedic, and heroic elements. Was weather modification a military boondoggle? an unconscionable misuse of public funds? a political and diplomatic embarrassment? a scientific dead end? a way of making money by taking advantage of drought-stricken farmers? Although weather control possesses some of these attributes, and no serious weapons were developed during the Cold War, it remains under discussion today as a special fantasy of military planners. Weather modification's intellectual offspring, cloud physics, has developed into a distinguished specialty niche among atmospheric scientists. In the private sector, routine cloud seeding by private companies, although providing at best a 10 percent enhancement of precipitation, is still a flourishing business, especially in the American West. Moreover, purposeful climate modification, as a response to inadvertent climate change, remains a speculative, and to some a terrifying possibility.

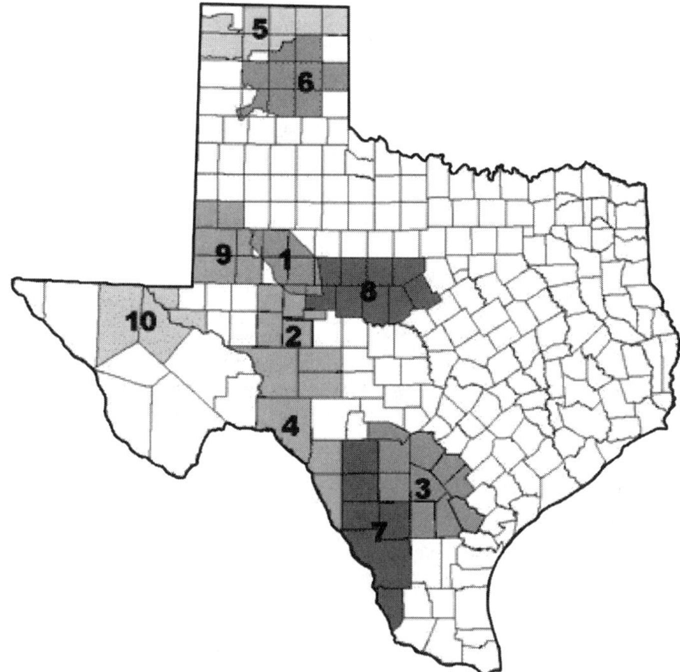

Fig. 9.2 Texas Weather Modification Programs as of December 2002 cover approximately one-third of the area of the state. The Texas Department of Agriculture provided funding of $2.4 million for cloud seeding activities in 2002 – equivalent to about 7 to 9 cents per acre. Areas with ongoing weather modification programs (and date of founding) include: 1. Colorado River (1971); 2. West Texas (1996); 3. South Texas (1997); 4. Texas Border (1998); 5. North Plains (2000); 6. Panhandle (2000); 7. South West Texas (1999); 8. West Central Texas (2001); 9. Southern Ogallala Aquiver Rain-Enhancement (2002); 10. Trans-Pecos (2002). Courtesy of Texas Department of Agriculture http://www.agr.state.tx.us/iga/grants_funding/iga_weathermod.htm (28 May 2003). Similar projects are ongoing throughout the American West. See, for example, Nevada, http://cloudseeding.dri.edu/HomePage.html (28 May 2003).

The next time a line of thunderstorms rumbles out of West Texas across the Dallas-Fort Worth area, or a hurricane veers offshore, or a sporting event or military operation is favored with good weather, perhaps you will remember the story of military and civilian schemes for cloud seeding and climate engineering. Perhaps your local weather or the climate of your region, continent, or the globe is far from natural. Perhaps someone has already "fixed" it.

Notes

1. C. G. Rossby, Chair, Panel on Meteorology, Joint Research and Development Board, Department of Defense, "Interim Report on Artificially Induced Precipitation," April 21, 1947, cited in secret memo, L. R. Hafstad to the Joint Staff, February 5, 1948, Subject-Numeric Files 1947–1953, 100-GG-GAT--Weather Control, Defense Research and Development Board, Office of Secretary of Defense, RG 330, National Archives and Records Administration (hereafter: RDB Weather Control Files, NA). In the same file see also confidential memo, C. G. Rossby, chairman of Panel on Meteorology to Committee of Geophysical Sciences, Joint Research and Development Board, n.d. but ca. January 1947; and secret memo, D. B. Langmuir, director, Planning Division to chairman, Committee on Geophysical Sciences, February 21, 1947.
2. See confidential memo, P. F. Lee, chief of Naval Research to assistant secretary of the Navy for Air, January 14, 1948, File CD 3-1-46, Cloud Modification Experiments, Office of the Secretary of Defense, September 1947–June 1950, RG 330, National Archives and Records Administration (hereafter: OSD Cloud Modification File, NA). On the U.S. Army's interest see confidential memo, "Process of Artificial Production of Precipitation," John L. Pfeiffer to director of Research and Development, War Department General Staff, September 17, 1947, RDB Weather Control Files, NA.
3. In support of this assertion, see Charles C. Bates and John F. Fuller, *America's Weather Warriors, 1814–1985* (College Station: Texas A&M University Press, 1986), 52 and passim. Early rainmaking, including its military connections is treated in Clark C. Spence, *The Rainmakers: American "Pluviculture" to World War II* (Lincoln: University of Nebraska Press, 1980). James P. Espy, employed by the army and the navy in the 1840s and 1850s was the first federal employee to advocate weather modification; on Espy's military connections see James Rodger Fleming, *Meteorology in America, 1800–1870* (Baltimore, MD: The Johns Hopkins University Press, 1990).
4. Col. Donald Yates, AWS commander in 1946 is quoted as saying, "I'm tired of consultants. In the next war, we're going to have our own damned Ph.D.'s already in uniform." He encouraged promising young officers to apply for permanent military commissions, and indicated that assignments to leading universities for graduate study were likely; Donald Yates to Charles Bates, January 29, 1946, cited in Bates and Fuller, *Weather Warriors*, 134.
5. See Bates and Fuller, *Weather Warriors*, 134 and 299 (note 6). The authors claimed that 80 percent of U.S. *civilian* meteorologists still had military ties. The numbers cited are as follows: 5,273 total, 2,267 in uniform, 1,325 U.S. Air Force reserve, 650 naval reserve, 1,031 civilians. This means that 4,242 individuals, or 80 percent of *all* U.S. meteorologists in 1954, had military ties.

    Naming names, Robert M. White, a lieutenant in the wartime U.S. Army Air Forces who received his doctorate through the GI Bill, went on to high-level administrative work as head of the Air Force Cambridge Research Center's Meteorology Development Lab from 1952 to 1959. He was also chief of the Weather Bureau, first director of ESSA (Environmental Science Services Administration), then NOAA (National Oceanic and Atmospheric Administration, and "meteorological czar" of the interagency Office of the Federal Coordinator for Meteorological Services and Supporting Research. In 1983 he was elected president of the National Academy of Engineering. The three immediate past directors of the National Weather Service, George P. Cressman, Richard E. Hallgren and Elbert (Joe) Friday, were all former military meteorologists. The list could go on almost indefinitely: Brig. Gen. Kenneth C. Spengler (AWS reserve) served as executive secretary of the American Meteorological Society from 1946 to 1988; Hallgren is now executive director emeritus of the society. U.S. Air Force General and former director of AWS Albert J. Kaehn, Jr. was the society's president in 1987; and Norman A. Phillips, who served as a lieutenant in the AWS, became a pioneer in numerical weather prediction.

    ESSA was established in the Department of Commerce in 1965 by a consolidation of the Coast and Geodetic Survey and the Weather Bureau. ESSA was renamed the National Oceanic and Atmospheric Administration (NOAA) in 1970.
6. National Academy of Sciences, *Research and Education on Meteorology: An Interim Report of the Committee on Meteorology,* January 25, 1958 (Washington, DC, 1958).

7. Vincent Schaefer, "The Production of Ice Crystals in a Cloud of Supercooled Water Droplets," *Science* 104 (1946): 459. See Horace Byers, "History of Weather Modification," in Wilmot N. Hess (ed.), *Weather and Climate Modification* (New York: Wiley, 1974), 9.

8. "Project Cirrus — The Story of Cloud Seeding," G.E. Review (November 1952), 12.

9. During World War II the National Defense Research Council, the Office of Scientific Research and Development, the Chemical Warfare Service, the secretary of war, and the U.S. Army Air Forces sponsored research on gas mask filters, screening smokes, precipitation static, and aircraft icing studies. Langmuir and Shaefer were involved in this work. See Irving Langmuir, "The Growth of Particles in Smokes and Clouds and the Production of Snow from Supercooled Clouds," *Proceedings of the American Philosophical Society* 92 (1948): 167.

10. Irving Langmuir, "Summary of Results Thus Far Obtained in Artificial Nucleation of Clouds," in *Final Report: Project Cirrus*, G.E. Report No. RL-140 (Schenectady, N.Y., 1948), 18.

11. *General Electric 56th Annual Report and Yearbook* (Schenectady, N.Y., 1947), 27.

12. James R. Fleming, "Irving Langmuir's Weather Modification Experiments for General Electric and Project Cirrus: A Case Study of the Limits of Private Support for Science," paper presented at the annual meeting of the American Meteorological Society, 1984.

13. Irving Langmuir, "The Production of Rain by a Chain Reaction in Cumulus Clouds at Temperatures above Freezing," *Journal of Meteorology* 5, no. 5 (1948): 110ff. This theory considers the development of precipitation in warm clouds by the collision-coalescence of drops that grow so large they break up into smaller drops and are carried upward to repeat the cycle, forming a so-called chain reaction. Although accounts of these developments exist in the scientific and journalistic literature, until now they have not attracted the attention of professional historians of science and technology. See, for example, Byers, "History of Weather Modification," 3-24; and Barrington S. Havens (comp.), "History of Project Cirrus," *General Electric Research Laboratory Report No. RL-756* (Schenectady, N.Y., 1952). Archival materials are found in the Irving Langmuir Papers in the Library of Congress, and in the archives of the General Electric Company at the Schenectady Museum and in the archives of the State University of New York (SUNY) Albany.

14. Eli Goldston, "Legal Entanglements for the Rain-maker," *Case and Comment* 54, no. 1 (1949): 3–6.

15. A contract was signed by the Army Signal Corps and General Electric Company on February 24, 1947. The Office of Naval Research participated in the program on an equal basis under government order NA ONR 19-47. The U.S. Air Force furnished airplanes and support personnel for the project. "Interim Report on Artificially Induced Precipitation," RDB Weather Control Files, NA.

16. Harvard Law School Record, "Many Legal Entanglements Forecast for Man in New Role as Rain-Maker," clipping in File CD 3-1-46, RG 330, NA.

17. "Project Cirrus — The Story of Cloud Seeding," 13.

18. *New York Times,* January 13, 1949, 25.

19. Earl G. Droessler, "Federal Government Activities in Weather Modification and Related Cloud Physics," in *Final Report of the United States Advisory Committee on Weather Control*, Vol. 2, Howard T. Orville, chairman (Washington, DC, 1957), 253.

20. F. W. Reichelderfer, "General Statement of the Chief of the Weather Bureau with Reference to H.R. 4582," Hearings — Committee on Interstate and Foreign Commerce, March 18, 1948, in File CD 3-1-46, RG 330, NA. See also memo "Air Force Activities in Cloud Physics Research Program," D. C. Doubleday to Office of the Air Judge Advocate, March 31, 1948, and other documents in ibid.; also Byers, "History of Weather Modification," 16–17.

21. Irving Langmuir, "A Seven-day Periodicity in Weather in United States during April 1950," *Bulletin of the American Meteorological Society* 31 (1950): 386ff.

22. Cited in *Weathering the Storm: Sverre Petterssen, the D-Day Forecast and the Rise of Modern Meteorology*, ed. James Rodger Fleming (Boston: American Meteorological Society, 2001), 294.

23. Ibid., 294–95. Sverre Petterssen, et. al., *Cloud and Weather Modification: A Group of Field Experiments* (Boston, MA: American Meteorological Society, 1957), *Meteorological Monographs* 2, No. 11. The composition of the Advisory Group was as follows: E. R. Biel, climatologist, Rutgers University; C. L. Critchfield, physicist, Convair; S. Petterssen, chairman, meteorologist, University of Chicago; A. Spilhaus, meteorologist, University of Minnesota; H. J. Stewart, aerodynamicist, California Institute of Technology; A. T. Waterman, physicist, National Science Foundation; M. A. Woodbury, statistician, University of Pennsylvania; Thomas S. Malone, special assistant.

24. Droessler, "Federal Government Activities," 253–54; Byers, "History of Weather Modification," 25–27; and Special Steering Group on Artificial Cloud Nucleation, "Report of Aspects of Artificial Cloud Nucleation Program," MS, n.d., but ca. December 1952, RDB Weather Control Files, NA.

25. Byer's associates in the cloud physics group were Roscoe R. Braham Jr., Louis J. Battan, James P. Lodge, Guy Goyer, and James E. McDonald. See Roger A. Prior, confidential memo, "Report on Aspects of the Air Force ACN Project at the University of Chicago," January 22, 1953, RDB Weather Control Files, NA.

26. Vonnegut is mentioned briefly in E. J. Kahn Jr., *The Problem Solvers: A History of Arthur D. Little, Inc.* (Boston: Little, Brown & Co., 1986), 203.

27. Petterssen et. al., "Cloud and Weather Modification"; see also Droessler, "Federal Government Activities," 254; and Byers, "History of Weather Modification," 28.

28. See "Excerpt from First Annual Report to the President and the Congress by the National Advisory Committee on Oceans and Atmosphere, dated June 30, 1972"; reprinted in Senate Committee on Foreign Relations, Subcommittee on Oceans and International Environment, *Prohibiting Military Weather Modification: Hearings on S.R. 281*, 92d Cong., 2d sess., 1972, 156 (hereafter: *Prohibiting Military Weather Modification*).

29. American Meteorological Society, "Statement on Weather Modification and Control," ca. 1950, Appendix III of "Second Report of the Special Committee on Cloud Physics," RDB Weather Control Files, NA.

30. J. Murray Mitchell Jr. to Luis de Florez, April 24, 1961 (copy), Harry Wexler Papers, Manuscript Division, Library of Congress (hereafter: Wexler Papers, LC).

31. See Robert D. Elliott, "Experience of the Private Sector," in Hess (ed.), *Weather and Climate Modification*, 45–89.

32. "City Flip-Flop on Rainmaking," *Daily News*, November 5, 1951 (clipping); H. Victor Crawford to John C. Morrissey, June 21, 1951; a bibliography on legal and historical aspects is included in Helmut E. Landsberg, "Memorandum for the Record — Briefing on Weather Control," November 5, 1951; these and related items are in RDB Weather Control Files, NA.

33. Victor Boesen, *Storm*, chapter 8. URL: http://www.weathersage.com/texts/boesen/chapter8.htm (30 May 2003).

34. Confidential memo, "The Legal Implications of Artificially Induced Precipitation," to the Secretaries of War and the Navy, n.d., but ca. 1950, RDB Weather Control Files, NA.

35. F. W. Reichelderfer, chief of the Weather Bureau to Helmut E. Landsberg, Joint Research and Development Board, July 30, 1947; Vannever Bush to James Forrestal, January 15, 1948; both RDB Weather Control Files, NA.

36. Kenneth M. Arenberg, et al., *Weather Modification: Past, Present and Future* (Weymouth, MA, 1954), 40.

37. "Weather Control," MS, n.d., but ca. 1950, RDB Weather Control Files, NA. Other bills included H.R. 4864, S. 798, H.R. 3672, H.R. 4887 and House Joint Resolution No. 56.

38. The committee was created by act of Congress, August 13, 1953 (67 Stat. 559), as amended July 9, 1956 (70 Stat. 509). Orville and four others were appointed by President Eisenhower on December 9, 1953, and were confirmed by the Senate on January 25, 1954. Six other members of the committee represented government agencies.

39. Public Law 85-510, dated July 10, 1958 directed NSF to initiate and support a program of study, research, and evaluation in the field of weather modification; the foundation did so until 1968 when Public Law 90-407 removed this role. See National Science Foundation, *First Annual Report on Weather Modification 1959*, NSF-60-24 (Washington, DC, 1960), and subsequent annual reports, 1960–68.

40. The analogy between hydrogen bombs and hurricanes is from R. H. Simpson and J. Simpson, "Why Experiment on Tropical Hurricanes?" *Transactions of the New York Academy of Sciences* 28, no. 8 (1966): 1045–62; reprinted in Geoffrey McBoyle (ed.), *Climate in Review* (Boston: Houghton Mifflin, 1973), 193–205. Table 9.1 is adapted from ibid., and from W. D. Sellers, *Physical Climatology* (Chicago: University of Chicago Press, 1965), 106. From 1962 until the early 1970s, the U.S. Navy was a cosponsor of Project STORMFURY, an attempt to dissipate hurricanes by seeding them. See H. E. Willoughby, et. al., "Project STORMFURY: A Scientific Chronicle, 1962–1983," *Bulletin of the American Meteorological Society* 66 (1985): 505–14.

41. *New York Times* (June 15, 1947) 46, 1; quoted in Bruce Bliven, "The Rainmakers," *Challenge Magazine* (May 1953): 45. For other general concerns of the era, see D. J. Ritchie, "Reds May Use Lightning as a Weapon," *Missiles and Rockets* 5, no. 35 (1959): 13–14.

42. The quotations are from an editorial by Arthur Krock, "An Inexpensive Start at Controlling the Weather," *New York Times* (March 23, 1961); copy in Wexler Papers. LC.

43. See confidential memos, "Cloud Seeding Ammunition Development Project," C. P. Vansant to R. C. Coupland, January 26, 1949; "Ammunition for Cloud Seeding," Edwin R. Patzing to Chief of Ordnance, March 23, 1949; "Minutes of Meeting on Cloud Seeding Ammunition," Frances L. Whedon to R. B. Simpson, March 28, 1949; all in RDB Weather Control Files, NA.

44. Howard T. Orville, "Weather Made to Order?" *Collier's* (May 28, 1954): 25–26, emphasis added.

45. *Ibid.*, 26; original emphasis preserved.

46. University Corporation for Atmospheric Research, Preliminary Plans for a National Institute for Atmospheric Research (1959); Elisabeth Lynn Hallgren, *The University Corporation for Atmospheric Research and the National Center for Atmospheric Research, 1960–1970: An Institutional History* (Boulder, CO, 1974).

47. Quoted in "The Weather Weapon: New Race with the Reds," *Newsweek* (January 13, 1958): 54. There was also concern about Chinese capabilities in this area; see Department of Commerce, "Chinese Communist Weather Control Experiments," USDOC 60-21921, 21 August 1959 (Washington, DC, 1960).

48. *Newsweek* (January 13, 1958): 54.

49. Henry G. Houghton, "Present Position and Future Possibilities of Weather Control," in *Final Report of the United States Advisory Committee on Weather Control*, Vol. 2, 288; also quoted in *Newsweek* (January 13, 1958): 54.

50. E. K. Fedorov, "Modification of Meteorological Processes," ibid., 401. On other Soviet weather and climate modification projects see Nikolay T. Zikeev and George A. Doumani (comps.), *Weather Modification in the Soviet Union, 1946–1966: A Select Annotated Bibliography* (Washington, D.C., 1967).

51. Albert E. Burke, "Influence of Man upon Nature — the Russian View: A Case Study," in William L. Thomas, Jr. (ed.), *Man's Role in Changing the Face of the Earth* (Chicago: University of Chicago Press, 1956), 1036 and 1049–50; see also John Maynard, *Russia in Flux* (New York: MacMillan Co., 1948), 14–15.

52. George S. Benton, "Some General Comments on Meteorological and Weather Modification Activities in the Soviet Union," *Bulletin of the American Meteorological Society* 50 (1969): 918–22; Hess (ed.), *Weather and Climate Modification*, 385. According to Louis J. Battan, in 1976 Soviet investments in cloud physics and weather modification research were "substantially greater" than in the United States. Battan, "Weather Modification in the Soviet Union – 1976," *Bulletin of the American Meteorological Society* 58 (1977): 19.

53. Gordon J. F. MacDonald, "How to Wreck the Environment," in Nigel Calder (ed.), *Unless Peace Comes* (New York: Viking, 1968), 181–205; reprinted in Senate Committee on Foreign Relations, Subcommittee on Oceans and International Environment, *Weather Modification: Hearings*, 93rd Cong., 2d sess., 1974, 55ff. (hereafter: *Weather Modification Hearings*. A similar analysis was conducted by J. O. Fletcher, *Managing Climatic Resources*, RAND Report No. P-4000-1) (Santa Monica, CA: RAND Corporation, 1969).

54. Gordon J. F. MacDonald, statement in House Committee on International Relations, Subcommittee on International Organizations, *Prohibition of Weather Modification as a Weapon of War: Hearings on H.R. 28*, 94th Cong., 1st sess., 1975, 3 (hereafter: *Prohibition of Weather Modification*).

55. Jack Anderson, *Washington Post* (March 18, 1971) "Air Force Turns Rainmaker in Laos," F7. See also Seymour Hersh, "Rainmaking is Used as a Weapon by U.S … ," *New York Times* (July 3, 1972); reprinted in *Prohibiting Military Weather Modification*, 14; and Bates and Fuller, *Weather Warriors*, 229–32.

56. See memorandum from the deputy secretary of Defense to the Hon. Nicholas deB. Katzenbach, undersecretary of state, Subject: "Military Action Program for SE Asia," February 21, 1967; cited in Department of Defense, "United States-Vietnam Relations, 1945–1967: Study Prepared by the Department of Defense," Book 5, Vol. 2, *U.S. Ground Strategy and Force Deployments: 1965-1967* (Pentagon Press) (Washington, DC, 1971), 50–51. As early as 1957, then Senator Johnson had speculated on controlling the Earth's weather from space for military purposes; see Lowell Ponte, "Weather Warfare Forecast: Partly Cloudy — U.N. Treaty Would Permit 'Peaceful' Environmental Research by Military," *Los Angeles Times* (January 29, 1976); reprinted in Senate Committee on Foreign Relations, Subcommittee on Oceans and International Environment, *Prohibiting Hostile Use of Environmental Modification Techniques: Hearing*, 94th Cong, 2d sess., 1976.

57. *Weather Modification Hearings*, 87ff. The operational phase began on March 20, 1967, and was conducted each year during the monsoon season (March–November) until July 5, 1972. The project was also known as Intermediary Compatriot, and by AWS as Motorpool. See John F. Fuller, *Air Weather Service Support to the United States Army: Tet and the Decade After*, AWS Historical Study No. 8 (Scott AFB, Ill.: Military Airlift Command, 1979), 30–32.

58. Deborah Shapley, "Weather Warfare: Pentagon Concedes 7-Year Vietnam Effort," *Science* 184 (June 7, 1974); reprinted in House Committee on International Relations, Subcommittee on International Organizations and Movements, *Weather Modification as a Weapon of War: Hearing on H.R. 116 and 329*, 93rd Cong., 2d sess., 1974, 35 (hereafter: *Weather Modification as a Weapon.*)

59. Westmoreland, *A Soldier Reports* (New York: Doubleday, 1976), 342. See also Tad Szulc, *The Illusion of Peace: Foreign Policy in the Nixon-Kissinger Years* (New York: Viking, 1978), 39.

60. Fuller, *Air Weather Service Support to the United States Army*, 30-32.

61. Air Weather Service, Fourth Annual Survey Report on the Air Weather Service Weather Modification Program [FY 1971], *AWS Technical Report* 244 (Scott AFB, Ill.: Military Airlift Command, 1972).

62. Testimony of Melvin R. Laird before the Senate Foreign Relations Committee, April 18, 1972; cited in Weather Modification Hearings, 109–10; see also Daniel S. Greenberg, "Vietnam Rainmaking: A Chronicle of DoD's Snowjob," *Science and Government Report* 2 (May 1, 1972): 1–4.

63. Cristine Russell, "The Weather as a Secret Weapon: From Vietnam to Geneva," *Washington Star* (August 23, 1975); reprinted in *Prohibition of Weather Modification*, 47.

64. "French Fight Off Dienbienphu Blow," *New York Times* (April 23, 1954), 2; cited in Arenberg, et al., *Weather Modification: Past, Present and Future*, 39. Cloud seeding also may have been used in Korea, especially to clear cold fogs, see e.g., Helmut E. Landsberg to H. C. Schweinler, July 20, 1950, "Use of Cloud Seeding in the Korean War," RDB Weather Control Files, NA; and personal communication Earl G. Droessler to the author, December 14, 1990.

65. Hersh, "Rainmaking is Used as a Weapon by U.S …"

66. Lowell Ponte, *International Herald Tribune* (June 29, 1976): 2.

67. Gordon J. F. MacDonald, statement in *Prohibition of Weather Modification*, 5.

68. For a detailed account of American environmental war efforts and their effects in Indochina, see Stockholm International Peace Research Institute, *Ecological Consequences of the Second Indochina War* (Stockholm: Almquist & Wiksell International, 1976). For background see Ruth Russell, "The Nature of Military Impacts on the Environment," in Sierra Club, *Air, Water, Earth, Fire* (San Francisco: Sierra Club, 1974), 1–14.

69. Philip Handler to Claiborne Pell, July 25, 1972 in *Prohibiting Military Weather Modification*, 153.

70. This data is from the NSF annual report report on weather modification for 1962; data for 1969 to 1979 are taken from Department of Commerce, *National Weather Modification Policies and Programs: A Report to the President and the Congress* (Washington, DC, 1979).

71. The request was filed with the Freedom of Information Act and Security Review Office at the Pentagon on October 9, 1990; the "no record" response was dated November 21, 1990; W. M. McDonald to author, November 21, 1990, author's personal files.

72. Georg Breuer, *Weather Modification: Prospects and Problems* (Cambridge: Cambridge University Press, 1979); transl. by Hans Mörth of *Wetter Nach Wunsch?* (Stuttgart: Deutsche Verlanganstalt, 1976), 144–45.

73. S. Res. 281, 92nd Cong., 2d sess., 1972.

74. "Hearing on Senate Resolution 281," *Bulletin of the American Meteorological Society* 53 (1972): 1185–91.

75. *Congressional Record* (11 July 1973): 233303-05. As adopted by the 93rd Congress the resolution was designated S. Res. 71. See Lawrence Juda, "Negotiating a Treaty on Environmental Modification Warfare: The Convention on Environmental Warfare and Its Impact upon Arms Control Negotiations," *International Organization* 32 (autumn 1978); reprinted in Senate Committee on Foreign Relations, *Environmental Modification Techniques: Hearing*, 96th Cong., 1st sess., 1979, 24–29.

76. Senate Committee on Foreign Relations, *Environmental Modification Treaty: Hearings*, 95th Congress, 2d sess., 1978, 87ff.

77. "Joint Statement on Environmental Warfare," in *Weather Modification as a Weapon*, 11. The text of the statement is as follows: "The United States of America and the Union of Soviet Socialist Republics: Desiring to limit the potential danger to mankind from possible new means of warfare; Taking into consideration that scientific and technical advances in environmental fields, including climate modification, may open possibilities for using environmental modification techniques for military purposes; Recognizing that such use could have widespread, long-lasting, and severe effects harmful to human welfare; Recognizing also that proper utilization of scientific and technical advances could improve the interrelationship of man and nature; 1. Advocate the most effective measures possible to overcome the dangers of the use of environmental modification techniques for military purposes. 2. Have decided to hold a meeting of United States and Soviet representatives this year for the purpose of exploring this problem. 3. Have decided to discuss also what steps might be taken to bring about the measures referred to in paragraph 1." Moscow, July 3, 1974.

78. Stephen S. Rosenfeld, "Weather as a Weapon," *Washington Post* (February 28, 1975), and "Weather Warfare," *Washington Post* (June 26, 1975); both reprinted in *Prohibition of Weather Modification*, 40. Both news accounts are based on the statement of Dr. Edith Brown Weiss, Brookings Institution, in *Weather Modification as a Weapon*, 21.

79. A. Gromyko to the secretary general of the United Nations, August 7, 1974, in *Weather Modification as a Weapon*, 11–12.

80. Juda, "Negotiating a Treaty," 28.

81. *Ibid.*, 24–29.

82. United Nations, *Multilateral Treaties Deposited with the Secretary-General: Status as of 31 December 1982* (New York, 1983), 667. The Text of the U.N. Convention (A/RES/31/72) is reprinted as Appendix C in Congressional Research Service, *Weather Modification: Programs, Problems, Policy, and Potential,* 95th Cong., 2d sess., 1978, 510–13.

83. Department of Defense, "Statement on Position on Weather Modification," [1978], Appendix B in *Weather Modification: Programs, Problems, Policy, and Potential*, 509. The statement, from the Office of the Under Secretary of Defense for Research and Engineering, was provided by Col. Elbert W. (Joe) Friday, former director of the National Weather Service.

84. See Stanley A. Changnon Jr., "Weather Modification in 1972: Up or Down?" *Bulletin of the American Meteorological Society* 54 (1973): 642–46; "Hearing on Senate Resolution 281"; and Gordon J. F. MacDonald, statement in *Prohibition of Weather Modification*, 4.

85. *Climate Change 1992: The Supplementary Report to the IPCC Scientific Assessment*, ed. J. T. Houghton, B. A. Callander, and S. K. Varney (Cambridge: Cambridge University Press, 1992), 5; Intergovernmental Panel on Climate Change, "Summary for Policy Makers: The Science of Climate Change," Working Group 1, 1995, <http://www.unep.ch/ipcc/wg1.html> (December 5, 1996).

86. United Nations, Information Unit on Climate Change, "Are We Overlooking the Social and Political Implications of Climate Change?" <http://www.unep.ch/iucc/fs108.html> (September 26, 1996).

87. David Keith and Hadi Dowlatabadi, "A Serious Look at Geoengineering," *EOS, Transactions of the American Geophysical Union* 73 (1992): 289, 292–93.

88. Jerome Namias, "The Greenhouse Effect as a Symptom of Our Collective Angst," *Oceanus* 32 (summer 1989): 66.

89. Melinda Liu, "Rain Called on Account of Games," *Newsweek* (August 5, 2002); Amanda Riddle, "Powder Dries Up Fla. Thunderstorms," *AP News*, July 19, 2001; "Florida Inventor Believes He Can Suck the Power out of Hurricanes," http://web1.wsvn.com/news/articles_p/local/C21236/ (1 June 2003); Col. Tamzy J. House, et al., "Owning the Weather in 2025," *Air Force 2025* (August 1996) available at the following URL: http://www.au.af.mil/au/2025/volume3/chap15/v3c15-1.htm (1 June 2003).

IV
Fixing Business

10

The "Problem" of Computer-Computer Communication, 1995–2000: A Technological Fix?[1]

PAUL E. CERUZZI

Introduction

If there is one common thread to the many discussions of what is called a *technological fix*, it is that those who proposed it did not think far enough to anticipate the side effects, the "unintended consequences," (in Edward Tenner's phrase), or the true place for a technological innovation in the larger context of the society into which it is introduced.[2] When looking at the history of the invention and spread of the electronic digital computer, one finds many examples of optimistic projections of how the computer would solve any number of problems of society — projections that later on seem naive and obviously wrong. For example, in the 1950s a number of visionaries (not always engineers who were actually doing the work) spoke of the coming Utopia of full-time leisure brought on by automation: when "the push button age is already obsolete: the buttons now push themselves."[3] The invention of the personal computer in the late 1970s elicited a similar set of predictions from what I call "Digital Utopians." In both cases society was indeed transformed, even drastically, but the Utopia that was promised failed to appear.

What follows is a look at an innovation that was at once more modest than what I have just described, and at the same time more revolutionary. It was the development of computing technology — largely software — to solve one of the most vexing problems that faced computers in their early incarnations, namely their inability to share data with one another. If one defines the problem that way, an inability to share data seems inconsequential compared to the problems of unemployment, pollution, urban decay, and public education, which the Digital Utopians promised to solve. Yet it was a serious problem, and its solution ranks among the greatest advances in computing since the invention of the electronic digital computer itself in the 1940s. When in 1991 I was asked to write a survey article about trends in computing for the third edition of the *Encyclopedia of Computer Science and Technology*, I stated that the communication problem was one of the two most urgent needs to be addressed in the field.[4] But even as I was writing those words, the problem was being solved.

The solution came in three parts, two of them technical and the third social. The first was the invention of the Internet, which used a technique known as *packet switching* to allow messages to be passed from one computer to another regardless of the type of machines involved. The second (the social innovation) was the spread of this technique, invented and initially controlled by the U.S. Department of Defense, into public, general, and unrestricted use. The third was the development of software known as the World Wide Web, and with it a program called a *browser* to access it, which provided direct and simple access to data stored on computers worldwide. A number of studies have appeared about the development of the Internet and Web, and they will not be repeated here.[5] Those studies have not, however, been able to place this innovation in the context of how a technological fix sometimes leads to consequences that not only surprise but even repel and shock their creators. It is not because scholars have been unaware of these issues; it is more that we are still in the midst of the revolution brought on by the Web, and it is impossible to get enough perspective to see more than just the outlines of an historical structure. Impossible or not, it is nonetheless worth a try, and looking at the second part, the social innovation, may be a good place to start.

The Public Internet

> I don't think any of us know where this thing is going anymore, …
> but there's something exciting happening, and it's big.

> — William Wulf, May 1993[6]

Professor Wulf probably thought he was exaggerating. He wasn't. Not since Dorothy remarked that she and Toto were not in Kansas any more has such a momentous change been described with such understatement.

The Internet was once something that a few professors in academia or engineers in the computer industry knew about. Many of us can remember the day when we realized it was going to be something bigger. That happened to me on an evening in November 1997. I was riding in a chauffeured limousine on my way to speak before a group of high-level industry executives. The topic was the history of computing, and what insights, if any, the study of history could offer to chart the future. I had prepared some remarks about the history of the Internet, and how it would facilitate collaboration among scientists, humanists, and others among the intellectual and professional elite in the country. On the way to the hotel the limo passed by a brightly lit billboard, on which was plastered a huge image of Shaquille O'Neal who, I vaguely knew, was a basketball player. Parts of the billboard were extended with strips of plywood to accommodate his gangly arms and legs sticking out in all directions. The text of the advertisement consisted of one phrase: "www.Shaq.com."

By the time I got to the hotel I realized that my talk was obsolete. Until that night I had understood the Internet in a narrow historical context: of attempts to access computers remotely, to build air-defense and airline-reservation networks, to time-share mainframes, and to share expensive resources. Now the Internet was something else. It was no longer only a facet of computing technology; now it was part of entertainment, consumer spending, and popular culture. The Internet had fused computing with the mainstream of social life in America.

No single event, not even Shaquille O'Neal's decision to mount a personal Web page, turned this innovation from one direction to another. In hindsight one can easily say that the commercialization of the Internet was inevitable, as people often do when looking back on the confusing tangle of facts as they happened. In fact, such a transformation could not have occurred without jumping over a number of hurdles — social, political, and technical. The Internet jumped over the technical hurdles so easily that it is often not even acknowledged: its ability to evolve from handling a few thousand nodes linked by 56-kilobyte lines to millions of nodes linked by ever-faster satellite, microwave, and fiber-optic lines. It would be hard to find another technology that scaled so well. The Internet scaled because of its robust design, one that put most of the network activities not on the physical network itself, but on the computers and routers that were connected to it. Because these end-devices, in turn, grew in capability and speed (following Moore's Law), the network was able to grow by a factor of 1,000 in speed and one million in number of hosts from 1969 to 1996, without experiencing any severe disruptions.[7] Continued growth after 1996, plus increasing commercial use, have put incredible strains on the network, yet it continues to function, even if roughly on occasion.

The Acceptable Use Policy

The political hurdle was how to accommodate obvious commercial traffic on a network that was conceived and built by contracts let by the federal government. In the early 1980s the focus of networking shifted from the

Advanced Research Projects Agency (ARPA) to the National Science Foundation (NSF), which managed a network called NSFNet from 1988 through 1995. The NSF assumed responsibility for the Internet in 1990, and the original ARPANET was decommissioned (the military evolved its own, restricted networks). The NSF had to address the question of how to deal with commercial firms being connected and using the net. It responded with an Acceptable Use Policy, which read in part:

"NSF Backbone services are provided to support open research and education in and among U.S. research and instructional institutions, plus research arms of for-profit firms when engaged in open scholarly communication and research. Use for other purposes is not acceptable."[8]

The policy allowed "announcements of new products or activities … but not advertising of any kind." Thus it was all right for, say, IBM to use the Internet to disseminate technical information about one of its products, especially if that would help users connect that product to the Internet. It could announce the availability of a new PC but not announce a year-end sale on them. The line was not clear, but the NSF tried to draw it anyway. The policy further allowed "communication incidental to otherwise acceptable use, except for illegal or specifically unacceptable use." That implied that personal e-mail and even discussion groups were allowed, as long as they did not dominate the traffic to or from a particular site. "Extensive use for private or personal business" was specifically deemed unacceptable. Shaq would have to wait.

By 1992 the restrictions were lifted. Traffic on the Internet, already growing rapidly, grew even faster — from 1 trillion bytes a month in January 1992 to 10 trillion a month in 1994. Professor Wulf, quoted at the beginning of this section, was a former engineer at the Digital Equipment Corporation, on leave from an academic post at the University of Virginia, and at the time in charge of the NSF's networking program in the late 1980s. Like the others at the research-oriented federal agency, he looked at the growth of traffic on NSFnet with a mixture of fear and excitement. Scientific knowledge in general had been growing exponentially since the seventeenth century, but not at these rates. The NSF had to get off the train before it accelerated any faster. In 1995 the NSFnet was dissolved, and the NSF got out of the business of running a network and back to funding research. The Internet was privatized.

But how? The particulars of this transfer are murky. Some of the confusion comes from a claim made by Vice President Al Gore Jr., who people thought tried to claim responsibility for this transition, in an interview with Wolf Blitzer of CNN in March 1999. Gore did not claim that he "invented the Internet," as his critics charged, but that was how the press reported it. The exact words were, "During my service in the United States Congress, I took the initiative in creating the Internet." Gore's blunder did not help his candidacy.[9] As a seasoned politician he knew how the press could distort a story, but what he apparently did not understand was that the public has little understanding of, or tolerance for, the nuances that accompany the *invention* of any major technology.

Looking back on the whole brouhaha, it appears that Gore was trying to claim credit for easing the transition to public usage: a transition on which the subsequent "revolution" depended, and one that obviously required some sort of legislative action to effect. For a television series on the Internet, produced in 1998, Stephen Segaller claimed that the crucial legislation came not from Gore but from Congressman Rich Boucher of Virginia, who in June 1992 introduced an amendment to legislation that authorized the NSF to "support the development and use of computer networks which may be used substantially for purposes *in addition to* research and education in the sciences and engineering ..."[10] According to Segaller, when President George H. W. Bush signed the bill into law, it effectively ended the Acceptable Use Policy. That may have been the law that effected the change, but Gore, not Boucher, played a more important role.

Even Gore's critics admit that as a senator, before he became vice president, he was a fierce champion of federal support of computer networking. If he did not coin the phrase "Information Superhighway," he promoted the concept tirelessly and was responsible for bringing that phrase into common currency.[11] One curious aspect of his gaffe to the press was that no one reported that Gore, along with many others at the NSF and elsewhere, envisioned a future Internet that was nearly the opposite of how things turned out. To summarize briefly the complex and rapidly evolving series of events, Gore's vision was reminiscent of the earliest days of the ARPANET. He wanted the federal government to assist in building a high-speed network, called the National Research and Education Network (NREN), which would allow researchers to gain access to scarce resources, especially expensive supercomputers.[12] With that net in place, scientists all across the country could push the frontiers of physics, chemistry, and above all biomedical research. The NSF in turn would get out of the business of running and paying for a network, but would insist that the scientists themselves pay for whatever networking they needed. They could, of course, include those charges as part of the grant applications to the NSF or any other funding agency; and people assumed that telecommunications companies would build a physical plant to respond to this market. That did indeed happen. With the opening up of the Internet to commercial traffic, there was a land rush to build such facilities. (Too many companies jumped in, and the bubble burst in 2001.) Ultimately, the demand for access to supercomputers or other scarce resources was small compared to a demand for general Internet access on PCs and workstations for many applications, of which scientific research was in the minority. The impact of opening up networking to science was enormous; it only seems small in comparison to the other things that happened when the Internet was opened to public access.

While a senator in 1991, Gore proposed a bill to create what he called a National Information Infrastructure, which would formalize this transition. The essential parts of the High Performance Computing Act (the "Gore Bill") were debated through 1992, and a version was eventually passed. Meanwhile, Gore left the Senate and became vice president in January 1993.[13] As vice president he continued to champion Internet usage, insisting that federal agencies set up Web pages containing basic public information about who they were and what they did. The White House set up a Web page, www.whitehouse.gov, which among other things, showed a picture of the First Family's cat (Socks). When someone clicked on the cat's image, it meowed. That does not sound like much in the twenty-first century, but in the context of personal computing in the early 1990s it was a major advance.

Alexander Graham Bell thought the telephone would primarily be a business tool and was surprised to find people using it for idle chat. Thomas Edison did not envision his phonograph being used for music and entertainment. Likewise, the commercial use of the World Wide Web was not foreseen by the Internet's inventors (and it had many "inventors," certainly not a single individual like an Edison or Bell). Symbolic of these unanticipated consequences occurred when Web surfers typed in www.whitehouse.com instead of www.whitehouse.gov: they were taken to a site offering pornographic materials (for a fee). Pornography drove much of the commercialization of the Internet, just as it did the early days of video recording and of motion pictures.[14] In 1992, the number of registered dot-com sites was well behind dot-edu (although ahead of dot-gov), but by mid-decade the dot-com world overwhelmed all the others, so it was no surprise that people typed in that suffix when trying to reach the White House.[15]

Java: Adding "Sizzle" to the Web

The preceding discussion of how commercialism came to the Internet does not respect the distinction between the Internet and the public perception of the Internet, accessed through the World Wide Web. The two are different and that distinction should be understood. Commercial activities, almost exclusively, are done via the Web. The Hypertext Transfer Protocol (HTTP), the protocol invented by Tim Berners-Lee to retrieve information on the Web, overwhelms transactions that use FTP, Gopher, WAIS, or remote login, the protocols that one used before the Web was invented. It was not just the Web's invention that paved the way for commercial use, however. In the early 1990s, just as this phenomenon was starting, Bill Joy of SUN Microsystems recognized a need for a programming language that would fit the times. He was no more certain of where the Internet was going than anyone else, but he believed that existing languages were not up to the task. He spoke of a need for a language that retained the advances of C++, then rapidly gaining popularity, but with more of the low-level power of C or assembly language. He called on programmers to write a language that was, he said, "C-plus-plus-minus."[16] Beginning in 1991 James Gosling, along with a small team of other programmers at SUN, came up with a language, called Oak, that filled Joy's needs. With Joy's support, it was reworked, renamed *Java* and publicly announced in March 1995. At the precise moment that commercial uses were being allowed on the Internet, and as the World Wide Web made navigating easy, along came Java: a language that enabled Web page designers to put the "sizzle" in their offerings. As everyone knows, it is the "sizzle" not the "steak" that sells.

Java quickly became the means by which Web designers could give their pages animation, movement, and interactivity. It caught on because a program written in Java could run on nearly any computer, large or small, from any vendor that was connected. As with the underlying design of the Internet itself, Java took advantage of the growing power of the PCs and workstations to provide the translating capabilities, so that the Java programmer could simply "write it once, run it anywhere." It did that by using a notion that went back at least to the beginnings of personal computing. When the IBM personal computer was announced in 1981, customers had a choice of three operating systems. Microsoft's DOS was the one that everyone remembers, but also offered were Digital Research's CP/M and the UCSD (University of California, San Diego) "p-system." This third system, written in Pascal, translated code not into machine language but into code for an intermediate *pseudo* machine (hence the name), which in turn was compiled and executed. Why this extra layer of complexity? The reason was that by letting each computer manufacturer write its own p-code compiler, the operating system would run on any number of computers without modification.

The p-system never caught on. One reason was that the IBM PC (and its clones) quickly became a standard, and all those other machines that might have taken advantage of the p-system never established themselves. Another reason was that the p-system required a two-stage translation process, which made it unacceptably slow compared to MS-DOS. And programmers did not care for the Pascal language, though it was admired in the universities; they preferred the raw, close-to-the-metal code of MS-DOS.

A dozen years later — an eternity in "Internet time" — the situation had changed. The Web was now hosting machines of a great variety and size — there were even plans to provide Web access to televisions, handheld organizers, and cell phones. Suppliers of hardware and software, seeing how Web access boosted sales for them, did not mind writing the code to translate from the Java pseudo machine. Finally, the brute force of the Intel Pentium processor, now standard for personal computers, was enough to overcome the inefficiencies of a dual translation. Note that the inefficiency is still there and will always be there, when compared to a program tailored for a specific machine. But given the mix of processor power, compiler, and language design, and telecommunications speeds of the mid-1990s, it mattered less this time.

As Java caught on, it garnered media interest all out of proportion to what the language actually did, and that was unfortunate. Java's write-once, run-anywhere feature was heralded in the trade press as a way, not to do something interesting on the Web, but to break Microsoft's hold on personal computing software. If people could write programs on large servers, and have those programs sent to the desktop over the Internet, who needed to buy the Office suite from Microsoft? If people could write programs in Java, which any computer could run, who needed Windows? It was a variant of what people were saying about Netscape's Navigator, and in both cases they were wrong. Microsoft was not enthusiastic about Java's popularity, although they got a license from SUN to use it. SUN later claimed that Microsoft violated the agreements, and these arguments made their way into the courtroom along with those coming from Netscape. For the hapless Web surfer, Java was a mixed blessing: it provided sizzle, all right, but a lot of Web page designers used it to design sites that had little else. Waiting for a Java-heavy page to load through a slow telephone connection, surfers experienced the first evidence of gridlock on Al Gore's Information Superhighway. Java has yet to prove its worth in making "smart" appliances like toasters, thermostats, or even cell phones, but people with things to sell on the Internet embraced it immediately.

With all these pieces now in place, the dot-com bubble followed naturally. The U.S. computer industry had already seen several of these. In the late 1950s there was a host of companies offering cheap, magnetic drum-based computers for sale. During the "go-go" years of the late 1960s, new minicomputer and software service companies were starting up every day. The late 1970s and early 1980s saw the Altair and later IBM-compatible PCs at the low end. That paralleled the JAWS (just another workstation) phenomenon at the high end of computing. All those bubbles burst, though none with the ferocity of what happened beginning around March 2000. Although this most recent bubble is fresh in everyone's minds, it is worth recalling some details of its origins.

In July 1995 Jeff Bezos opened an online bookstore he called Amazon.com. By October it was processing 100 orders a day. According to the company, by July 2000, "a 100-order minute is common …" The company has yet to demonstrate profitability, although Bezos was named *Time* magazine's Person of the Year in December 1999.[17] In September 1995 Pierre Omidyar started an online auction service called Auction Web that he hoped would help his girlfriend trade Pez dispensers (or so he claimed; critics claim that this story was a fabrication). Auction Web grew into eBay, with seven million ongoing auctions in 2001, trading items from baseball cards to new, used, and vintage automobiles, to a Gulfstream jet (which sold for $4.9 million).[18] Unlike Amazon and many other commercial sites, eBay is profitable. It levies a small charge to the seller to list an item, and levies another charge based on the selling price if the item is sold. Amazon and eBay were among the few Web businesses (other than pornographers) that have a steady revenue stream. The myriad of sites that relied on advertising (their ads often driven by Java) did not fare as well; many were in financial trouble or out of business by the summer of 2001. At the time of this writing the dot-com collapse is still in force, so it is too early to tell who among the start-up companies will survive and what will emerge from all the turmoil. Besides eBay and Amazon, mentioned above, the Internet portal Yahoo! seems to have achieved bulk and stability, although like Amazon it has been unprofitable and does not have a clear future.

If there is any pattern to be found among those commercial Web sites that have survived, it is that they provide their patrons with a sense of community. They remain commercial sites at their core, but users get at least a vestigial sense of what ARPANET must have been like, or when people created online communities on the first Usenet sites. Besides Usenet and the online bulletin boards, one community stood out: the WELL (Whole Earth 'Lectronic Link) located in Sausalito, California. It was founded in late 1984 as an electronic version of the *Whole Earth Catalog*. Whole Earth founder Stewart Brand, who helped start the WELL and gave it its name, played a key role in fusing computer technology with countercultural values. Users paid a modest subscription fee, and it ran over the slow dial-up modem connections of the day. From its beginnings, and on to the present day (it is still alive), it was notable not just for its Bay-area countercultural flavor, but also for the quality of the postings. It attracted a large number of writers and remains focused on writing, not graphics or multimedia. It was the home for an especially active group of fans of the Grateful Dead, who traded information about the band. According to chronicles of the era, the traffic from Deadheads kept the WELL financially solvent through perilous times.[19]

Among the members of the WELL was Howard Rheingold, whose writings promoted the concept of *virtual communities*. Another was John Perry Barlow, who was not only a Deadhead but even wrote lyrics for the Dead (to be sure, in the band's later, baroque period). Barlow was a cofounder, with Mitch Kapor, of the Electronic Frontier Foundation, and in 1996 he posted on the Web his "Declaration of the Independence of Cyberspace," a rant against those in the commercial and government world who were trying to muzzle the burgeoning Internet. It opened with the words, "Governments of the Industrial World, you weary giants of flesh and steel, I come from Cyberspace, the new home of the Mind. On behalf of the future, I ask you of the past to leave us alone. You are not welcome among us. You have no sovereignty where we gather."[20]

This phenomenon should remind us that nontechnical factors, so strong at the invention of the personal computer, continue to shape the direction of computing. In the 1970s, executives at the big computer companies could not understand how the Apple II could ever be successful. In the 1990s it was Web companies, and the Web itself, that created a new commercial paradigm, at a time when industry leaders were touting things like 500-channel cable television, or video on demand.

On the Amazon site, customers (including a book's author) were encouraged to write reviews of a book. Even more ingenious, others could rate a review. A review that panned a book, which in turn was rated low by other members of Amazon's community, could thereby be judged accordingly. Some, including former Representative Newt Gingrich, became well known for the quantity and quality of their reviews. College professors found students writing research papers that referred to books they had neither borrowed, bought, nor read: they simply looked at the reviews on Amazon.com and summarized the contents.[21] Authors afflicted with writer's block found a new excuse not to write: several times a day they would check Amazon's site to see how well (or poorly) their existing books were selling. Likewise, eBay devised a system of customer feedback that ranked the credibility of auctioning items. A seller who was late delivering goods or who falsely advertised the condition of a collectible doll, would soon be chastised by the community. The result was to lower the rate of fraud in a field (auctions of used goods) where fraud is common. The customers did the work, and it is unlikely that eBay could have been effective in policing its site anyway. Other Web sites let users buy and review consumer items, and then let others rate the credibility of the reviewer. Still others provided advice by "experts," whose reputation and authority came not from their academic degree or position but from how well their advice has been rated by those who followed it in the past. How many of these sites will be able to convert that communal feeling into profitability is an open question, but those who did *not* offer that had trouble staying in business after 2001.

As PCs found their way onto the desks of corporations, they carried with them the spirit of the hackers who created the first of their kind. So too did the Web's commercialization carry the communal spirit of the first online communities. That spirit was not the same of course, and some who had been present at the creation of the Web were unhappy to see it become a mass phenomenon. Still, a small but critical piece of John Perry Barlow's declaration is embedded in the commercialized Web, and an understanding of this spirit, along with the more obvious factors of developing a sound business plan, etc., separated those few commercial sites that have so far survived the shake-out from those who have not.

Search Engines, Portals, and Communities

Besides the commercial sites mentioned above, another group of Web sites that emerged were search engines and portals: sites that help people navigate the Web and find information of interest to them. The most successful of these was Yahoo!, founded in 1994 as "Jerry's Guide to the World Wide Web," by Dave Filo and Jerry Yang, while students at Stanford University. As the Web grew, a number of search engines appeared. Most used some sort of automated program (*bot*, short for robot) to "crawl" through Web sites and extract keywords. Yahoo! used automation too, but its index was assembled by Jerry and Dave — personally at first, later supplemented by other people. So from the outset Yahoo! was more than just a search engine but a place where a human touch took some of the fear out of visiting the Web naked. The service was free. When in August 1995 the site began running advertisements to generate revenue, a few complained but the site stayed popular.[22] The site got an early boost from Netscape, which put a link to it from its own site. When people installed the Netscape browser, Netscape's home page was the first thing they saw, and many never bothered to change that. As Netscape's management was railing against Microsoft, the company did not realize, until it was too late, that it had also invented a portal but did not know it. Netscape was eventually acquired by AOL, another company that was adept in easing ordinary people into the complexities of cyberspace.

Computer-savvy Internet users did not need a portal. They preferred brute-force search engines and were not afraid to construct complex searches using Boolean algebra to find what they wanted. An early leader was Altavista, founded by the Silicon Valley lab of Digital Equipment Corporation. Altavista's success was not enough to rescue its parent company, and by late 1998 its search engine was passed by a rival, Google, founded by Sergei Brin and Larry Page (also of Silicon Valley). Google's success lay in the way it dealt with the retrieval of a keyword: it ranked sites containing that word according to how much that site was referenced by other sites, again taking advantage of the built-in community of cyberspace.

Google's success illustrates something fundamental about the Web: as it emerged by the end of the 1990s, the Web was flawed. Tim Berners-Lee, its creator, described how only half of his vision has come true; he initially wanted a Web that was easy to write to as it was to surf.[23] Each user should be able to construct his or her own portal tailored precisely to one's needs — a notion that Yahoo! has followed with its "My Yahoo!" feature but nowhere near as integral to the Web as Berners-Lee envisioned. Likewise, among the Web's harshest critics are those who the history books have credited as being pioneers in linked information structures: Ted Nelson, who coined the term *hypertext* for a Web-like system he called Xanadu, and Doug Engelbart, who created an "On Line System" and the device he called the *mouse* with which to navigate it. They argue, to an audience that increasingly does not want to listen, that users pay a high price for the flat, one-way file structure of the Hypertext Transfer Protocol.[24] Almost forgotten, too, was a system for the retrieval of scientific information developed by Eugene Garfield, called Science Citation Index. Like Google, it indexed scientific papers by noting how many other papers, if any, referenced them in their footnotes. Citation Index ran on traditional mainframes and made a transition to the Internet age, but it is no longer a major force in helping navigate thorough knowledge spaces. Brewster Kahle, the creator of a pre-Web indexing system called Wide Area Information Service (WAIS), also recognized the Web's shortcomings, and in 2001 he launched an Internet Archive, which allows users to retrieve old Web sites that have otherwise disappeared into the ether.[25]

Tragedy of the Commons

In 1968, at the onset of the modern-day environmental movement, the biologist Garrett Hardin wrote a paper called "The Tragedy of the Commons." It became widely cited and referred to in the decades that followed as one of the most insightful observations on the causes of the environmental problems afflicting industrial societies.[26] In the paper he likened global environmental problems to a common grazing area of a village, where the cost of letting "just one more" animal graze is low, while the benefit to the owner of that animal is high. When the total population is low, the damage to the commons is insignificant, but as the number of grazing animals increases the damage gradually becomes more severe, until the entire commons is destroyed. In the paper, Hardin quantifies the economic relationship of benefits to individuals versus damage to the common resource, but he also argues that nothing short of external regulation can prevent this tragedy from occurring. Neither unfettered market forces nor altruistic behavior by the common's users will suffice.

By a strict technical measure, the Internet has not come close to approaching this point of overpopulation. Many, including Bob Metcalfe, who predicted its collapse as it grew exponentially, have had to eat their words.[27] The Internet passed through challenges like the 1988 worm, viruses, the Y2K crisis, the dot-com collapse, and the terrorists' attacks of September 11, 2001 with hardly a hiccup. It is based on a robust design. As for the content and quality of information that the Internet conveys, however, it has indeed been tragic. Web surfing has gotten so frustrating that one has a hard time remembering how exhilarating it once was. The collapse of the dot-coms may have been a signal that something was wrong, like the collapse of fisheries in the North Atlantic Ocean. More recently, the proliferation of spam is threatening the utility and convenience of e-mail. More such collapses may be forthcoming. Regulation, which Hardin proposed for the natural environment, will not come easily to the Internet if people like John Perry Barlow have their say. Consolidation and control by a few giant media companies like AOL Time Warner may come to the fore. Neither future is appealing. A third solution is possible: leave the infrastructure of the Internet in place, but reconstruct its content space along the lines of what researchers like Nelson, Engelbart, and Berners-Lee have advocated.[28] That would not be an easy task, but neither was the construction of the Internet and the World Wide Web in the first place. This in an ongoing issue and it is too early to say how, or if, it will be resolved.

Conclusion

The technological fix worked. Information flows transparently across the World Wide Web among IBM mainframes, special purpose supercomputers, UNIX workstations, Microsoft Windows computers, Macintoshes, and other devices. By simply typing in a Uniform Resource Locator (URL), one can access information stored on machines physically located anywhere in the world. Programs written in Java allow a measure of interactivity that earlier generations of computer scientists hardly dreamed possible. A person can ask a question on Google, using no extensive training, no specialized terms or vocabulary, and get an accurate answer within seconds.[29] In the United States, an individual can purchase a computer capable of doing all this for less than $1,000, and be connected to the Web over his or her household telephone lines for a modest monthly fee. How, then, is one to respond to the well-founded criticisms of the Web as it exists today? Is it simply an illustration of the engineers' axiom that "the better is the enemy of the good": that a working system, with its inevitable flaws, is preferred to a Platonic, perfect system that never gets built? What about the argument that the Web has not brought us Utopia? After all, people like John Perry Barlow predicted that it would. With all that we know of the history of twentieth century technologies such as the airplane, nuclear power, or the mass-produced automobile, and of how they too promised a peaceful and harmonious world that failed to materialize, we surely ought to know not to ask such a question. Many (not Barlow) who predicted such a future for a Web-enabled world were blinded by the money being made during the dot-com bubble. Others knew the history of technology but felt that the patterns of, say, the history of the airplane did not apply to the development of cyberspace. Initially I was reluctant to come to an optimistic conclusion about the Web's recent history, but I am more reluctant to criticize a technology that I depend on so much in my daily life and work. Thanks in part to information we gather over the Web, we in the twenty-first century are aware, as those before us were not, of the dangers of looking for a technological fix for all our ills.

Notes

1. This paper is derived in part from a chapter of Paul Ceruzzi, *A History of Modern Computing*, 2nd ed. (Cambridge: MIT Press, 2003).
2. Edward Tenner, *Why Things Bite Back: Technology and the Revenge of Unintended Consequences* (New York: Knopf, 1996).

3. John Diebold, "Factories without Men: New Industrial Revolution," *The Nation* (1953): 227–228, 250–251, 271–272.

4. The other need was ease of use, which to this date has yet to be solved. Paul Ceruzzi, "Digital Computers: History," in Anthony Ralston and Edwin D. Reilly, eds., *Encyclopedia of Computer Science*, 3d ed. (New York: Van Nostrand Reinhold, 1993), p. 459.

5. The basic histories are Janet M. Abbate, *Inventing the Internet,* (Cambridge, MA: MIT Press, 1999, Tim Berners-Lee, *Weaving the Web,* (San Francisco: Harper San Francisco, 199), and Jim Clark, with Owen Edwards, *Netscape Time:* (New York: St. Martins Press, 1999).

6. Wulf, quoted in Christopher Anderson, "The Rocky Road to a Data Highway," *Science* 260 (21 May 1993), pp. 1064–5.

7. Moore's Law, named after a cofounder of the Intel Corporation, Gordon Moore, states that the density of circuits on a computer chip, and by implication computing power, doubles every eighteen months and has done so ever since the invention of the chip in 1959. The best statement of the Internet's underlying design philosophy is found in the unsigned "request for Comment" (RFC 1958: "Architectural Principles of the Internet," found at www.ietf.org/rfc/rfc1958.txt.) This was written in June 1996. After 1996 the crush of commercial traffic put strains on the design that it has managed to handle but with greater difficulty. See also the National Research Council, Computer Science and Telecommunications Board, *The Internet's Coming of Age* (Washington, DC, National Academy Press, 2001) Chapter One.

8. The Acceptable Use Policy can be found on several Web sites, as well as reprinted in Ed Krol, *The Whole Internet User's Guide & Catalog* (Sebastopol, CA: O'Reilly & Associates, 1992), pp. 353–354.

9. Since he lost by a few hundred votes (the margin in Florida), one should say that any number of other missteps that he made, not just this one, cost him the election. The text of his interview can be found on several Web sites, among them the "Urban Legends" site, which I have relied on for this discussion.

10. Emphasis mine. Boucher's language is quoted in Stephen Segaller, *Nerds 2.0.1: A Brief History of the Internet* (New York: TV Books, 1998), p. 296. The book was the companion to the television series broadcast on the PBS network.

11. See, for example, an article he wrote for the September 1991 Special Issue of *Scientific American* on "Communications, Computers, and Networks," (September 1991, pp. 150–-153); also a speech he gave at UCLA on January 11, 1994 to the "Superhighway Summit," which may be found at http://artcontext.com/cal97/superhig.txt, accessed December 13, 2001.

12. In the mid-1980s C. Gordon Bell, formerly Digital Equipment's chief engineer, was detailed to the NSF, where he championed the notion of supercomputer centers and may have been the key architect of the NREN plan.

13. Christopher Anderson, "The Rocky Road to a Data Highway," *Science* 260 (21 May 1993), pp. 1064–1065.

14. Jonathan Coopersmith, "Pornography, Technology, and Progress," *ICON* 4 (1998), pp. 94–125.

15. Peter H. Salus, *Casting the Net* (Reading, MA: Addison-Wesley, 1995), p. 222. Salus cites figures for January 1992, of 243,020 registered dot-edu domains vs. 181,361 dot-com, and 46,463 dot-gov domains.

16. Joy used this term both privately to SUN programmers and in public speeches, including one given at the National Air and Space Museum in 1990 attended by the author.

17. Historical information was taken from Amazon's web site, accessed July 2000.

18. Robert D. Hof, "The People's Company," *Business Week*, (e-biz section), December 3, 2001, pp. EB-15–-21.

19. Stephen Segaller, *Nerds 2.0.1* (New York, 1998), p. 269. Also Katie Hafner, *The Well: a Story of Love, Death & Real Life in the Seminal Online Community* (New York: 2001).

20. The text can be found at the EFF's Web Site: www.eff.org, accessed in November 2001. Barlow has his own Home Page, which can be accessed through the EFF's site, although it appears that it has not been updated recently.

21. The author has done some of the research for this essay in this manner.

22. Robert H. Reid, *Architects of the Web: 1,000 Days that Built the Future of Business* (New York: John Wiley, 1997), Chapter 6.

23. Tim Berners-Lee, *Weaving the Web* (New York: Harper Collins, 1999).

24. For Ted Nelson's home page, see http://www.sfc.keio.ac.jp/~ted/ (accessed December 2001).

25. See www.isinet.com/isi/products/citation/sci/ (accessed December 2001); Brewster Kahle's site was founded in 2001 at www.archive.org.

26. Garrett Hardin, "The Tragedy of the Commons," *Science* 162 (13 December 1968), pp. 1243–8.

27. Metcalfe literally ate the page of the magazine in which he predicted the collapse of the Internet due to increased traffic.

28. See for example, Nathaniel S. Borenstein, "The Once and Future Internet," paper presented at a Symposium on the occasion of the tenth anniversary of the first U.S. Web site, at the Stanford Linear Accelerator Center (SLAC), December 3–4, 2001. An outline may be found at http://www-project.slac.stanford.edu/webanniv/nsb.pdf

29. The computer scientist Richard Gabriel has argued that Google has in fact, passed the "Turing Test" for Artificial Intelligence: that a person typing in a question to Google usually gets an answer that is as good as or better than the answer he or she would get by asking a human expert.

11

Innovation Junctions

ONNO DE WIT, JAN VAN DEN ENDE, JOHAN SCHOT,
and ELLEN VAN OOST

One striking aspect of the twentieth and twenty-first centuries is the rise of a number of organizationally and geographically distinct spaces — cities, factories, households, hospitals, harbors, supermarkets, airports, offices, to name some of them — as important sites for technology development. As the century progressed, the number of different technologies in simultaneous use in these spaces increased. This collocation of technologies encouraged various actors to develop mechanisms and arrangements by which they could coordinate the interaction of these technologies. Typically these actors also developed mediating technologies that facilitated and stimulated the interaction of different technologies.

A number of scholars have reflected on the management activities required to deal with the complex interactions of collocated technologies. Joel Tarr and Gabriel Dupuy have shown how, at the end of the nineteenth century, as the growth of Western cities spawned serious logistical challenges concerning energy and water supplies, waste handling, transportation, and communications, municipal bureaucrats responded by beginning to coordinate and plan the layout of urban areas.[1] Lindy Biggs has presented a picture of a new kind of industrial engineer starting to systematically plan the physical design of factories and develop assembly lines and conveyors to handle the flow of materials between machines and work stations in the factory.[2] In fact, the concepts of *urban technology* and *factory technology* gained currency by the increased collocation of technologies at specific sites. As the city and factory shaped technologies, so in turn did technologies shape the city and the factory.

219

In this article we analyze the collocation of technologies and the resultant patterns of development created in a specific site: the office. For this purpose we advance the concept of the *innovation junction*, which we define as a space in which different sets of heterogeneous technologies are mobilized in support of social and economic activities, and in which, as a result of their collocation, interactions and exchanges among these technologies occur. These interactions and exchanges lead to location-specific innovation patterns. The problems posed and opportunities offered by the collocation and interaction of different sets of technologies in bounded spaces create a need for coordination. This need is defined not only by users themselves (companies, managers, employees) but also by a new type of intermediate actor positioned between producers and users of technologies, working and reflecting on their interaction. Like the city and the factory, the office became the subject of intense analysis and intervention by these reflexive actors. Two forms of interaction among artifacts resulted as well: the combined use of two or more technologies and the extension of the functional characteristics of technologies, including the transfer of functional characteristics from one technology to another.

Innovation junctions are not exclusively twentieth-century phenomena; many factory and harbor technologies (to name only two possibilities) were applied concurrently in earlier eras. However, in the twentieth century, innovation junctions became increasingly important to the development of technology and society. Their impact on twentieth-century society is comparable to that of large, geographically dispersed, infrastructural systems, such as electrical grids, communication networks, and transportation systems.[3] They led to the emergence of a new range of infrastructures, products, activities, services, and industries, and to new sets of user patterns and identities. However, innovation junctions differ in fundamental ways from the infrastructure-based systems that have been studied using the large technical systems perspective inspired by Thomas Hughes's classic *Networks of Power* (1988).[4] Hughes's study offers a generalized model of the process of large system development, arguing that a number of characteristic phases in that development can be distinguished: invention, development and innovation, transfer, growth, and momentum. During each phase, specific groups are foregrounded — engineers-entrepreneurs, manager-entrepreneurs, and finally financier-entrepreneurs — to produce solutions to the problems encountered in building the system. In addition, in Hughes's model system, development proceeds through a number of mechanisms: attention to a favorable load factor and economic mix, identification of reverse salients, and the creation of momentum.

Although understanding the mechanisms of systems development may be important for understanding the development of innovation junctions, the Hughesian framework is not appropriate for analyzing them. The main reason for this is that the change dynamics of innovation junctions are not determined by the process of creating and sustaining geographically dispersed, material infrastructures. Instead, technology development at innovation junctions takes place by dealing with the challenge posed by the existence of different sets of technologies at one location. Whereas large technical systems derive their dynamics from geographical expansion, innovation junctions derive their dynamics from collocation in specific spaces.[5] Their local and spatial base also distinguishes innovation junctions from second-order large technical systems. The concept of second-order large technical systems has recently been introduced as a necessary complement to the first-order large technical systems analyzed by Hughes.[6] Second-order large technical systems — for instance, postal systems or organ transplant systems — are built on top of first-order large technical systems, combining and utilizing elements of them. In this sense, second-order large technical systems are similar to innovation junctions, which also incorporate different sets of technologies. However, in contrast to innovation junctions, second-order technical systems are highly dispersed. As a result they do not foster a location-based interaction and exchange among sets of technologies. Also, although first- and second-order large technical systems contain sites of technology development, these are loosely or tightly coupled through large-scale infrastructural networks. Thus, airports and harbors can be viewed as hierarchically connected sites within larger systems of transportation. Factories, households, and offices, in contrast, are far less systematically coupled, although they too are part of common sociotechnical networks.[7]

Ruth Schwartz Cowan has pointed to the importance of user location in explaining technological development. In her view, to understand technical change, and especially diffusion, one needs to identify the consumption-production networks (ranging from producers through intermediate actors to final consumers) and to view those networks in which new technologies are produced from the consumer's point of view.[8] Cowan defines the household as "the consumption junction, the place and the time at which the consumer makes choices between competing technologies."[9] The way we address junctions is inspired by her concept, especially her emphasis on users, but differs from it in a fundamental way. We will emphasize innovation patterns that result from the combined use of different technologies and the transfer of functional characteristics between technologies. Cowan's analysis is restricted to competing technologies and consumer choices, while our focus is on problems and opportunities raised by the collocation of a number of technologies.[10]

The Office as Innovation Junction

Our focus here is the development of the office in the Netherlands between 1880 and 1980, roughly the period between the introduction of the typewriter and the introduction of the personal computer. Since the kind of interactions we were looking for took place on the leading edge of office technology development and were therefore first implemented in larger offices, we have left smaller offices out of our analysis. New research will have to establish whether the technology dynamics identified in this article also hold true for smaller offices, offices outside the Netherlands, and other kinds of innovation junctions.[11] Our focus is not so much on what actually went on in the office — at the shop-floor level, so to speak — as on the design of the processes of interaction. Accordingly, we have primarily studied the actors responsible for decisions concerning the procurement and application of office technology.[12] We will draw mainly on case material from a number of Dutch companies in the service sector and in trade and industry, and on sales literature from office machine producers and importers.

The office as a distinct domain started to emerge in the Netherlands in the second half of the nineteenth century. At about the same time, copying presses and hectographs were introduced to supplement the traditional pen and paper. Large offices appeared at the end of the nineteenth century and continued to grow during the first decades of the twentieth century. A host of new technologies surfaced, among them typewriters; accounting, bookkeeping, stencil, and addressing machines; and punched-card installations. In the 1920s and 1930s, offices with over a thousand employees and substantially mechanized administrative operations were not unusual.[13] Computers followed in the 1950s. These technologies were employed in the performance of a number of different activities, ranging from the production, copying, storage, and transmission of documents to bookkeeping and accounting to administrative data processing. Since all were used in the same location and for more or less interconnected tasks, they had to be geared to each other to an increasing extent and so became, in various ways, more and more interdependent.

The existing literature on the history of the office and the history of office technologies can be roughly divided into three main categories. One category describes and analyzes the development of office technology in the context of changing labor and gender relationships.[14] Another focuses on specific user locations at which specific technologies, such as punched-card machines or computers, were introduced and applied.[15] A third considers the design of office machines and technologies at the level of specific artifacts.[16] Generally speaking, this literature as a whole does not systematically analyze the processes of interaction inside the office either at the level of the dynamics of office technologies or at the level of the actors involved. One notable and inspiring exception is JoAnne Yates, whose *Control through Communication* (1989) offers on the one hand, historical overviews of several early office technologies and on the other, case histories of the ways in which internal communication processes were transformed at several major American companies.[17] However, in contrast to the approach taken in this article, Yates does not conceptualize these interaction processes inside the office, and limits her analysis to text production, copying, and filing.

This article discerns three phases in the history of the office, each characterized by specific patterns of interaction between office machines and producers, users, and intermediary actors. We will discuss each of these periods in turn before proceeding finally to an analysis of the patterns that resulted from those interactions and a discussion of how they differ from the dynamics of large technical systems.

New Office Machines, 1880–1914

Between 1880 and 1914, the development of new management practices, the rise of large firms, and the establishment of new government requirements spurred demand by owners and managers of companies for new methods of bookkeeping and cost accounting.[18] Several new technologies accompanied the new office methods, and among these the typewriter was of central importance. The American Remington Typewriter Company began selling typewriters in Europe, including the Netherlands, in the early 1880s, and the trading company Fred. Stieltjes and Company became Remington's Dutch sales representative in about 1883.[19] By shortly after the turn of the century, one major office equipment company, Blikman and Sartorius, was selling almost two hundred typewriters a year.[20]

Available data suggest that in the Netherlands as elsewhere the first users of the typewriter were predominantly top male managers.[21] However, the typewriter was very soon transferred to the domain of male or female secretaries, who combined typewriting with dictation, shorthand, and telephone work.[22] Because the typewriter speeded up the process of text production, managers gradually separated the conceptualization and writing of text.[23] Innovations that were introduced almost simultaneously with the typewriter, such as the dictating machine and various shorthand writing techniques, enabled this separation.[24] Typewriter agencies capitalized on this development by combining typewriter courses with courses in shorthand writing.[25]

Typewriting also came to be linked to the reproduction of texts and documents. Before the introduction of the typewriter, the production and copying of a text were two separate activities: after the text was written, it was copied with the help of a copying press or hectograph. With the invention of carbon paper suitable for typewriters, it became possible to combine writing with copying.[26] At the turn of the century, almost all Dutch typewriter ads promoted this possibility.[27] With carbon paper, up to twenty copies of a letter or invoice could be made.[28] Also, typewriting was compatible with a new copying technology introduced during the 1890s, stencil duplicating. Although cyclostyles or mimeographs, as stencil duplicators were called in the United States, at first were used to make copies of handwritten texts, after the introduction of the typewriter stencil paper was developed that was suitable for making typewritten stencils. A 1904 advertisement claimed that the combination of the mimeograph and the national typewriter could produce as many as two thousand readable copies.[29]

The increased mechanization of the office not only affected the production and reproduction of letters and documents, it also influenced their filing and storage. The introduction of typewriters and carbon paper provoked the introduction of alternatives to the bound ledger. Duplication of current account notes, for example, became unnecessary if a loose-leaf current account book was used in combination with the typewriter and carbon paper. Files or ledgers consisting of loose leaves also made it feasible to arrange documents thematically instead of in the traditional chronological scheme. In some instances, the relationship between typewriter and file systems was reversed: at a number of insurance companies and registry offices (local census bureaus), new card systems stimulated the introduction of typewriters.[30]

Thus, although new office technologies were not necessarily introduced simultaneously at individual companies, the presence of the typewriter made it much more likely that changes in duplicating methods and file systems would soon follow. The typewriter became the center of a new administrative organization in which the technologies employed in producing, reproducing, and storing documents were increasingly linked. Although these technologies were discrete in principal, several forms of interaction emerged during this period at local offices and out of local initiatives. Sales agents and importers of office technologies in turn picked up on these interactions, pointing to the efficiencies created by the combined use of new technologies to reinforce their sales efforts.

System Machines and Efficiency Experts, 1914–1955

In the 1910s, accountants, efficiency experts, business managers, senior public servants, management consultants, and psychologists hotly debated new ideas of systematic and scientific management emanating from the United States. They argued for rationalized and more efficient procedures and management in a wide range of municipal, national, and industrial companies and service sectors. Through the implementation of office technologies and the systematic analysis of information flows and work procedures, managers and outside consultants also sought to optimize the monitoring and control of administrative activities and financial operations.[31] In the 1920s their individual activities became institutionalized as part of a broader efficiency movement. Members of this movement increasingly directed their attention to office management and administrative work procedures as opposed to production. Several consulting firms and institutions were established that came to mediate between the producers and importers of office machines on the one hand and the users of these office machines on the other.

In 1922 the Dutch Mechanical Administrative Office Organization (Mechanisch-Administratieve Bedrijfs Organisatie, MABO) came into being, founded by a number of employees from the Amsterdam headquarters of the Rotterdamsche Bank, who sought to commercially exploit the various reorganization and mechanization measures that they had been developing since 1917.[32] This was not the first Dutch organizational consultancy, but it was the first to concentrate exclusively on administration and office technology.[33] Partly in competition with the private MABO, beginning in about 1928, the Governmental Office Machine Center (Rijkskantoormachinecentrale, or KMC) started advising government departments on the purchase of office machines. One of the tasks given to the KMC was to assess the ways in which new office machines fitted into the existing office organization.[34]

Of the many organizations and institutions founded in the 1920s to promote efficiency in businesses and households, the Dutch Institute for Efficiency (Nederlands Instituut voor Efficiency, NIVE) was undoubtedly among the most important.[35] Founded in 1925, it established various "management research groups" that studied administrative methods and office machines, visited companies, and set up inquiries into the diffusion of office technologies. Beginning in 1927, the Research Group for Modern Office Technology (Studiekring voor Moderne Kantoortechniek, STUMOKA) also served as an important medium for transferring experience with office technology from one company to another. Members included a number of large banks, insurance companies, public utilities, and industrial companies. In 1930 STUMOKA officially became part of NIVE.[36]

The ad hoc implementation of office machinery during the years before World War I thus gave way to a much more systematic and professional approach. Whereas between 1880 and 1914 companies had simply experimented with new technologies, sometimes in cooperation with sales agents and importers, they now could turn for advice to management consulting firms and other intermediary organizations focused exclusively on rationalizing administrative activities and introducing office equipment. Office machinery was subsumed in a larger concern for efficiency, reorganization, and control.

The emergence of these intermediary organizations was closely related to technological developments. Most of the office machines that had been introduced during the previous phase now underwent a broad process of diffusion. Typewriters became common in the Dutch office, although certain types of offices, such as municipal registries, lagged behind. A 1926 handbook on office machines mentioned no fewer than twenty different brands of typewriters. Its author rightly remarked that companies without typewriters had become almost unthinkable.[37]

Beginning about 1914, as the typewriter and counting and adding devices became broadly diffused in the Netherlands, various multifunctional office machines appeared that could perform simultaneously a number of previously separate administrative activities. Examples of these "system machines," as they came to be called in the 1950s, include typewriting cash registers and typewriting counting machines as well as bookkeeping, punched-card, and addressing machines.[38] As the number and complexity of office machines increased, the demand for advice, instruction, and information also rose. This demand was still met in part by the importers and suppliers of office machines. In fact, during this period they institutionalized and professionalized their instructional activities.[39] But management consulting firms increasingly came to meet the demand for information on how office machinery and reorganization could be used to optimize administrative processes.

Addressing machines, introduced in the Netherlands just before World War I, undoubtedly became the most versatile system machines. Basically, these consisted of a punching device or typewriter that stamped addresses into metal or stencil sheets that were then linked together on a mimeograph machine. Companies that periodically had to send letters to large numbers of clients were the first purchasers, but beginning in the 1920s addressing machines came increasingly to be used for broader administrative purposes, such as producing lists of client data or printing names and addresses on invoices, bills, tickets, patients' identity discs, and notices. To make all this possible, addressing machines were fitted with mechanisms that supported list management and selection, printing, and tabulating. Because of their increased multifunctionality, it became easier to link addressing machines to other office machines, such as bookkeeping and punched-card machines. The practice at the Dutch central circulation bank, the Nederlandsche Bank, which was one of the first firms to purchase an addressing machine, offers an example. The printed lists produced by the bank's punched-card installation were cut into standard sizes in letter format. An Addressograph and a signature machine, the Graphotyp, then processed the lists into bills to be sent to the bank's clients.[40]

Increasing multifunctionality also allowed addressing machines to be developed into *organization systems*, a term first used in the 1920s and 1930s by the Andrema Company, an early manufacturer. The term was intended to differentiate between machines used for ordinary addressing work and more complex models around which the entire administrative process could be organized.[41] In some companies addressing machines became the central building blocks of the administrative organization. In the 1930s, for instance, local Dutch health insurance companies used punched identity discs for patients' forms, doctors' and pharmacists' cards, and internal administration registers. This application of addressing machine technology was codeveloped by the insurance companies and the supplier of the Andrema addressing machine, H. A. Kramers & Zoon.[42] Similarly, in 1920 a number of health insurance companies went to

Kramers & Zoon looking for an addressing machine that could print lists itself; the company began development, and the new machine was patented by Andrema in 1923.[43]

Punched-card devices also became important system machines. Herman Hollerith had developed these machines at the end of the nineteenth century in the United States for statistical applications, and they were first used on a wide scale in the 1890 U.S. census.[44] In Hollerith's system, clerks used keypunches to record data onto cards that were subsequently processed by sorting and tabulating machines. Punching, sorting, and tabulating machines could not operate independently, and in fact worked in close conjunction with each other. In addition to the Hollerith Company, which in the 1910s became part of the Computing-Tabulating-Recording Corporation (CTR), the Accounting Machine Company, established in 1911 by James Power, became an important manufacturer of punched-card machines. These two companies competed for the American and European markets, and both set up local Dutch agencies around 1920.[45]

Punched-card machines were introduced into the Netherlands in 1916 by the Dutch Central Statistical Office (CBS), which used the Hollerith machines to calculate foreign trade statistics. The numbers were read off from the tabulating machine and copied manually onto Burroughs adding machines.[46] In the 1920s the number and types of companies that deployed punched-card machines had increased significantly.[47] Among this new group of users were insurance and industrial companies, municipal services, state-owned enterprises, and banks. All had extensive administrative operations, which justified the acquisition of the costly punched-card equipment. The Rotterdamsche Bank was one of the main Dutch pioneers in the application of punched-card technology, and came to serve as a model for other large companies, in particular for the Dutch Giro Service and the Dutch State Mines. Both companies reorganized administratively, with the help of the former bank employees who in 1922 had established the MABO, along the lines set up at the bank.[48]

At first the Rotterdamsche Bank's punched-card installation was used only in the bank's giro transfer system: for every transfer a punched card was made, and these cards were then sorted, tabulated, checked against the earlier totals, and finally used in the making up of transfer notes. By 1921 some three thousand transfers a day were being handled in this manner.[49] In the years that followed, the number of Hollerith and Powers punched-card machines at the Rotterdamsche Bank's Amsterdam office increased significantly, as the machines were deployed for an increasing variety of administrative purposes.[50] This became evident during a 1926 exhibition on office organization and technology held in Amsterdam.[51] The exhibit catalog showed that calculating, bookkeeping, and punched-card machines had become central to the processing of all financial data, not just giro transactions. The bank was using punched-card machines to make up daily balance sheets for central management and to keep salary records, and the technology had been adopted in the bank's forwarding and postage department as well.[52]

The large-scale application of office machines at the Rotterdamsche Bank was accompanied by a systematic analysis of administrative processes aimed at optimizing them to suit information flows within the bank and identifying strategic points at which machines could be used to speed up work processes. Such analyses were common.[53] In the meantime, a number of innovations in the 1920s and 1930s allowed punched-card technology, which had until then been used primarily for statistical applications, to be adapted to bookkeeping and even more general administrative tasks. Improvements in methods of adding, subtracting, and multiplying helped to speed up statistical calculations and made it possible to perform more complex types of calculations. Tabulators were introduced that could print alphabetical characters in addition to numbers, an important innovation because it allowed names and addresses to be printed as well as numerical data, thus greatly enhancing the amount of information that could be provided in list form or in other sorts of documents.[54]

Both printing and alphabetical capability were innovations pioneered by the Powers' Accounting Machine Company. Powers produced an alphabetical tabulating machine as early as 1924, some seven or eight years before its rival CTR (which in 1924 changed its name to the International Business Machines Corporation) came up with a similar device.[55] The Powers alphabetical tabulator was introduced in the Netherlands in 1926.[56] It was quickly adopted by the Rotterdamsche Bank and by some of the other big Dutch banks for their securities administration. In 1932 the Twentsche Bank became the first European company to use an IBM alphabetical tabulator.[57]

Not all of the technical innovations in punched-card technology came from outside the Netherlands. At the Rotterdamsche Bank, Jan Stuivenberg, who had been involved in the design and application of punched-card machinery as early as 1920, played an important role in the successful adaptation of punched-card technology. Stuivenberg introduced several improvements to the bank's Hollerith and Powers machinery, including an automatic paper feeder for the Powers tabulator.[58]

Especially after the Second World War, producers and users of typewriters and system machines such as addressing machines and punched-card machines began to devote increasing attention to methods of exchanging information among them. Devices were developed that aimed at optimizing the transfer of data between typewriters and system machines. In all cases, punched cards and punched paper tapes served as interfaces. The Flexowriter, for instance, was an electric typewriter with an additional perforator that automatically transferred written text onto punched paper tapes. The tape could be used to operate the card punch of an addressing machine, or punched cards could be created from it using a tape-to-card converter. The process worked the other way around as well; several machines, including the Flexowriter, the Cardatype, and the Justowriter, could process punched tapes or cards into written text.[59]

The Flexowriter and Justowriter came onto the market in the beginning of the 1950s.[60] One of the companies that developed such devices in the Netherlands was the Philips Electronics Company. Stock movements were registered and billed by a small bookkeeping machine. Simultaneously, a card punch connected to the bookkeeping machine produced a punched card, which was subsequently processed by a tabulating machine. By connecting bookkeeping and punched card technology in this way, Philips eliminated several intermediate operations. In 1955 the connecting device was put into use (for Philips' stock administration).[61]

Research into integration devices continued at Philips during the second half of the 1950s as researchers studied methods of determining differences between written and punched data and of converting punched cards to punched paper tapes or vice versa. As more and more typewriters and bookkeeping machines produced punched paper tapes, the question of which medium was most appropriate as an integration device, punched cards or punched tapes, also arose. Because tapes were suitable for data transmission and because the similarity between written and punched data could more easily be checked in tapes, Philips settled on them.[62]

To summarize, after about 1914 the introduction and application of office machines grew more complex. The successful integration of system machines into an administrative organization called for much closer cooperation between suppliers and users, who also frequently called on consulting firms for advice about office technology. Sometimes these consulting firms were users who sought to transfer their own experience with office technology to other locations, emphasizing the interdependencies among efficiency, new management styles, and office machines. Devices with different functions, such as text production and data processing, came increasingly to be linked to each other: addressing machines were used in combination with typewriters or punched-card machines, punched-card machines in conjunction with calculating machines and printers. Both the addressing machine and the punched-card machine developed from rather simple technologies into complex data processing units, drawing together a number of administrative activities — counting, sorting, writing, listing, and the like — that had previously been separate. In addition to this functional integration, punched cards and tapes developed into media for transferring data between different office technologies. As a result, office technology increasingly came to mean a chain of stand-alone devices integrated by means of punched cards and punched tapes.

Integrated Data Processing, 1955–1980

From the mid-1950s on, the electronic computer gradually changed the Dutch office landscape. Although the computer was developed for scientific calculations, it did not take long for office equipment companies as well as office managers and efficiency consultants to recognize its potential as a data processing technology in the office. The 1953 introduction by IBM of the relatively small and cheap IBM 650 marked the definitive breakthrough of computers in the office; an IBM 650 installed in 1957 by the Dutch Land Cultivation Company (Nederlandse Heidemaatschappij) was the first computer in the Netherlands specifically used for office automation.[63] Although the company had never been in the forefront of office technology, its involvement in numerous postwar reconstruction projects led to a dramatic and increasingly problematic growth in administrative operations.[64]

Technically, at the Heidemaatschappij and elsewhere, the computer was a novelty. The first generation of computers employed vacuum tubes, which had already been used in punched-card tabulating machines such as the IBM 604, which had appeared on the market in 1948.[65] The crucial difference between electromechanical punched-card machines and computers lay in the fact that computers had internal processing units and memories that could store not only data but also programs.[66] This made the computer much more programmable than punched-card machines. It also demanded different programming techniques.[67] Internal processors and random access memory gave computers the capacity to perform long series of operations in a single step. As a result, the computer integrated functions (sorting, collating, tabulating, and the like) for which punched-card installations needed different machines. Finally, computers were much faster than punched-card machines. Despite the fact that early computers were expensive, unreliable, and difficult to use, these new features were attractive enough for several actors to invest in domesticating this promising technology into an efficient office machine.[68] These actors embraced the computer as a symbol of the modern office and as a forceful instrument of the schematization and systematization of office work.[69]

A continuing intensive interaction among hardware suppliers, efficiency engineers, office managers, and employees accompanied the implementation of the computer in the office. Despite this pattern of interaction, users tended to become dependent upon the programming and systems knowledge of hardware suppliers. Since the programming of first generation computers was system dependent and programming techniques differed from those of punched-card installations, office managers and efficiency engineers could only partially rely on their earlier experiences. But managers and engineers were uncomfortable with the dominant position that their lack of knowledge gave hardware suppliers, and they responded by using existing intermediary organizations, such as NIVE and STUMOKA, to exchange knowledge and experience about office automation. In 1958 they established a new institution, the Study Center for Administrative Automation (Stichting Studiecentrum

Administratieve Automatisering, SSAA). The SSAA, through various working and research groups, played a pivotal role in the development of office automation in the Netherlands until the mid-1970s. It became the central forum for managers, management consultants, economists, engineers, and mathematicians who were, in one way or another, involved with the growing field of information processing. As such, the SSAA became the cradle of a nascent group of computer professionals. It also quite successfully strengthened the position of user organizations with respect to hardware suppliers.[70]

Initially the computer constituted only a minor break as far as the interaction among office machines was concerned. Punched cards and punched tapes continued to serve as interfaces between computers and other devices. The Rotterdamsche Bank, for instance, introduced in the early 1960s a system to handle counter transactions by bookkeeping machines that produced a punched tape. During the night a tape-fed central computer processed all transactions.[71] In general, in this period punched cards and tapes only increased in significance, as they were the main input media for the first- and second-generation computers.[72] Significantly, when IBM introduced its third-generation 360/40 computer, which was suitable for both computing and data processing, the company called it a punched-card installation inside a computer.[73] In practice, punched-card departments were gradually transformed into central computing departments.[74] Often this transformation was accompanied by the integration of departments that had previously existed alongside each other.[75]

Although computers could process large sets of data faster than mechanical machines, the technology for producing the data input hardly changed. This led to an almost insatiable demand for typists, and it was difficult to staff the noisy punching rooms.[76] Several technologies were developed to automate data entry, not only by computer suppliers like IBM but also by technical departments in user organizations.[77] A relatively simple innovation was the so-called dual-purpose card, which contained both handwritten and machine-readable information. Probably the best-known dual-purpose cards in the Netherlands were remittance forms in the shape of a punched card; introduced by the Dutch Giro Service in 1961, when the service began to automate its account administration, these remained in use until the middle of the 1980s.[78] Dual-purpose cards could integrate the administrations of various actors — companies, banks, account holders, and clients.[79] An energy company, for instance, could send a punched card to a client, who would then manually fill in the missing information, keep one part of the card for their records, and then send the card to the Giro Service, where it would be automatically processed.[80]

In general, the tendency toward the integration of more tasks into a single machine that dated from before the computer era intensified between 1955 and 1980. Experts writing on office technology now explicitly began to call the new way of working *integrated data processing*, thereby also retroactively labeling the tendency toward integration in the first half of the 1950s with this term.[81] Particularly in third-generation computers, which possessed time-sharing and multitasking capabilities and could be controlled from terminals or consoles, interaction with other office machines, such as Flexowriters and Justowriters, became largely superfluous. The computer simply took over the tasks of devices such as addressing machines. Remote control and data transfer over telephone lines replaced the traditional cards and tapes. Multipurpose databases could be used simultaneously for different purposes by different departments within the same organization. In the early 1970s hardware suppliers indeed developed several types of remote desktops (keyboard and monitor) to be used for administrative processes.[82] In the second half of the 1970s these remote workstations began to be equipped with processing units and connected to electronic typewriters or printers, thus creating the possibility of integrating data processing and text processing.[83] These so-called multiuser systems were similar in functionality to the personal computers that would conquer the office desktop in the 1980s.[84]

The computer continued the trajectory of punched-card machinery in the office but accelerated the pace of the extension of functional characteristics. This was particularly true of the third-generation computers that appeared in the 1960s, which had better systems for handling databases and possessed time-sharing capabilities. The computer assumed the functions of several other machines and office technologies, and data transfer between computers and other devices became far less important. This meant that some existing interfaces between different office technologies became superfluous and were, in fact, incorporated in the design of the computer.

The Dynamics of the Innovation Junction

This article has attempted to analyze the workings of the office as an innovation junction at two different but connected levels: at the level of artifacts, and at the level of emerging and changing relations between producers and users of office technology. We argue that the office evolved through a set of specific interactions between artifacts and actors. The office that finally came about was partly an unintended consequence of this process, but also partly the deliberate aim of a number of reflexive actors pushing for the development of the modern office.

At the level of artifacts, the development and use of one type of office technology interacted with the development and use of other types. Two forms of interaction can be distinguished, both of which, in different periods and in different combinations, occurred in the office. First, two or more technologies could be used in combination. Often this combined use was accompanied by innovation in one or more technologies or by the creation of interfaces between one or more technologies. Examples of combined use include the typewriter with carbon paper and Dictaphone and the extension of the function of the typewriter through the interface of punched tape. Other examples are the Flexowriter and Justowriter; these had only a relatively short life before they were made superfluous by the second form of interaction, the functional integration of the digital computer. Second, the functional characteristics of technologies could be extended, or the characteristics of one technology transferred to another, leading to functional integration. Functional integration may refer to different technologies being used concurrently in one location, one of which takes over functions of human beings or of other technologies, as punched-card machines took over functions of bookkeeping and text-producing technologies. Functional integration may also refer to a new technology integrating the functions of different old ones, as the digital computer in a short period of time took over functions of various existing devices (punched-card machines, typewriters, bookkeeping machines, and desk calculators).

Generally speaking, the functional characteristics of an office machine and technology were at first clearly related to a specific functional domain inside the office: the production, reproduction, filing, or archiving of documents, as well as data processing, computing, or communication. Typewriters, for instance, were intimately linked to the production of letters and other documents, whereas punched-card machines primarily served data processing tasks. At first, interaction mainly took place between machines and technologies inside a specific domain, such as document production, reproduction, and filing. However, interaction increasingly crossed the borders of the various office activities.

The interaction process between different technologies was accompanied by historical changes in the relations among the actors involved. Four different actor groups can be distinguished: the producers of office machines and computer software, which in the Dutch case were almost without exception foreign companies; Dutch importers of office machines and office machine dealers, who often were represented in one and the same company; professional groups involved in the introduction and application of office technology, particularly efficiency engineers and, later on, computer professionals; and the users of office machines. It is important to note that the third group did not yet exist when the modern office started to emerge at the end of the nineteenth century. This group later became crucially important in creating a network of offices that are not hierarchically connected, but evolve interactively.

Based on the interaction processes and the various roles played by these actors, it is possible to distinguish three phases in the history of the Dutch office, each characterized by specific types of interaction between office machines and actors. In the first phase, one of ad hoc integration, which lasted from about the 1880s until World War I, several new office machines were introduced that mechanized only one isolated administrative activity, such as writing or counting. One of these machines, the typewriter, was at the center of several other technologies applied to the production and reproduction of documents. In this period, three groups of actors were active in the office — producers, importers, and users — but none was dominant.

During the second period, the era of partial integration, lasting from about 1914 until about 1955, new kinds of more complex office machines were introduced. We discussed in particular addressing machines, which mechanized the process of sending large numbers of letters, and punched-card machines, which process large volumes of numerical (and, later, alphabetical) data by sorting and counting. These different types of system machines, as this new generation of office technologies was called in the 1950s, increasingly combined the mechanization of several administrative activities — for instance, data processing and printing. Moreover, punched cards and punched paper tapes served as interfaces between system machines and a wide range of stand-alone office machines. The introduction and use of these office technologies often resulted in dramatic changes in office organization. More than in the previous phase, office managers had to adapt administrative processes to the machines. Large-scale service industries created separate departments for processing punched cards. During this period consultants came to form a new and important professional group, which pushed unceasingly for integration. They advised on the organization of the office, including the introduction and implementation of new technology.

During the period from about 1955 until about 1980, a period of centralized integration, a transition took place to integrated data processing and electronic data processing. For some companies, the introduction of computers for administrative purposes, which in the Netherlands started in the mid-1950s, was a revolutionary break from the past. For others, computer technology replaced punched-card devices and some other office machines. Punched cards and tapes remained important in the early part of the computer age, and in fact became the main input media for computers. During this phase the position of users and intermediary professional consultants was threatened by hardware producers and their national representatives. However, users and their consultants were able to regain independence through the creation of new platforms for knowledge transfer, such as the Study Center for Administrative Automation.

Of course, further research is needed to determine whether this phase model is specific to the history of the Dutch office or can be generalized. It also remains to be seen whether and to what extent the model is appropriate to the study of other innovation junctions, such as factories and households. Based on the history of the Dutch office, this article has tried to demonstrate that the dynamics of technological development in innovation junctions are different from those operating in large technical systems. Developments at innovation junctions do not follow a trajectory of invention, development and innovation, transfer, growth, and momentum. Rather, this case showed a process of ad hoc, partial, and centralized integration. Further, the change dynamics at innovation junctions are mainly dominated by interaction processes among users, various kinds of intermediate actors, and producers, including importers. It is not possible to identify system builders, although intermediate actors, such as efficiency engineers and management consultants, were important in identifying needs and opportunities for integration. Mechanisms such as load factors, economic mix, and reverse salients are not easily applied to the study of technical change at innovation junctions. Instead, a driving force must be located in the systematic effort to combine and connect various kinds of office machines. This effort has been provoked and reinforced by the continuing process of introduction of more machines at a bounded space, the collocation of various technologies, and the emergence of a new ideal of systemization and rationalization leading to the ideal of the rational office.

Notes

1. Joel Tarr and Gabriel Dupuy, eds., *Technology and the Rise of the Networked City in Europe and America* (Philadelphia, 1988).
2. Lindy Biggs, "The Engineered Factory," *Technology and Culture* 36, suppl. (1995): S174–88; *The Rational Factory: Architecture, Technology and Work in America's Age of Mass Production* (Baltimore, 1996).
3. On large technical systems, see Renate Mayntz and Thomas P. Hughes, eds., *The Development of Large Technical Systems* (Frankfurt am Main and Boulder, Colo., 1988); Thomas P. Hughes, *American Genesis: A Century of Invention and Technological Enthusiasm* (New York, 1989); Todd R. La Porte, ed., *Social Responses to Large Technical Systems* (Dordrecht and Boston, 1991); Renate Mayntz, "Grosse technische Systeme und ihre gesellschaftstheoretische Bedeutung," *Kölner Zeitschrift für Soziologie und Socialpsychologie* 45 (1993): 97–108; Jane Summerton, ed., *Changing Large Technical Systems* (Boulder, Colo., 1994); M. Karlsson and L. Sturesson, eds., *The World's Largest Machine: Global Telecommunications and the Human Condition* (Stockholm, 1995); Arne Kaijser and Marika Hedin, eds., *Nordic Energy Systems: Historical Perspectives and Current Issues* (Canton, Mass., 1995); Olivier Coutard, ed., *The Governance of Large Technical Systems* (New York, 1999).
4. Thomas P. Hughes, *Networks of Power: Electrification in Western Society, 1880–1930* (Baltimore, 1988).
5. This, of course, does not mean that innovation junctions do not expand.
6. Ingo Braun and Bernward Joerges, "How to Reconcile Large Technical Systems: The Case of European Organ Transplantation," in Summerton, 25–51.
7. Braun and Joerges refer to firms and households as the opposite ends of the continuum of large technical systems, with second-order systems positioned halfway between these two poles; Braun and Joerges, 44–45.
8. Ruth Schwartz Cowan, "The Consumption Junction: A Proposal for Research Strategies in the Sociology of Technology," in *The Social Construction of Technological Systems: New Directions in the Sociology and History of Technology*, ed. Wiebe E. Bijker, Trevor Pinch, and Thomas P. Hughes (Cambridge, Mass., 1987), 261–80. See also Thomas J. Misa, *A Nation of Steel: The Making of Modern America, 1865–1925* (Baltimore, 1995); Misa argues convincingly that innovation in steel manufacturing can only be understood by focusing on changing user-producer relationships.
9. Cowan, 263.

10. The issue of the interaction of technologies has been addressed from different viewpoints. Economic historians such as Nathan Rosenberg have pointed to the importance of interdependencies or complementarities among different technologies in explaining economic growth; see "Technological Interdependence in the American Economy," *Technology and Culture* 20 (1979): 25–50. Space has become an important concept in economics, especially as it contributes to understanding why innovations (defined as successfully introduced new products) are often clustered. Such a clustering delivers competitive advantages through spillovers, learning effects, informal information sharing, and scale effects. See Peter Hall, *Cities in Civilization* (New York, 1998), 291–500. See also John Seely Brown and Paul Duguid, *The Social Life of Information* (Boston, 2000); using the example of Silicon Valley, they argue that space remains meaningful in the age of globalization and expanding information flows. Francesca Bray, in *Technology and Gender: Fabrics of Power in Late Imperial China* (Berkeley and Los Angeles, 1997), has stressed the importance of studying different sets of technologies simultaneously. To understand the construction of gender, the relations between people, and the lack of overt change in China, Bray analyzed various technologies that provide overlapping contexts for the processes of social construction. We share Bray's desire to leave single-technology case studies behind and her focus on incorporating a spatial dimension in the history of technology, but we add an emphasis on analyzing the shaping role of collocation of various technologies, which cannot be found in her work.

11. This research is well underway as part of the Dutch national research project on the history of technology in the twentieth century.

12. This choice, of course, resembles Hughes's focus on system builders.

13. Onno de Wit and Jan van den Ende, "The Emergence of a New Regime: Business Management and Office Mechanization in the Dutch Financial Sector in the 1920s," *Business History* 42 (April 2000): 87–118.

14. See, for example, Alan Delgado, *The Enormous File: A Social History of the Office* (London, 1979); Margery W. Davies, *Woman's Place Is at the Typewriter: Office Work and Office Workers, 1870–1930* (Philadelphia, 1982); Joli Jensen, "Using the Typewriter: Secretaries, Reporters and Authors, 1880–1930," *Technology in Society* 10 (1988): 255–66; Sharon Hartman Strom, *Beyond the Typewriter: Gender, Class, and the Origins of Modern Office Work, 1900–1930* (Urbana, Ill., 1992); Angel Kowlek-Folland, *Engendering Business: Men and Women in the Corporate Office, 1870–1930* (Baltimore, 1994); Francisca de Haan, *Gender and the Politics of Office Work: The Netherlands, 1860–1940* (Amsterdam, 1998).

15. Arthur L. Norberg, "High-Technology Calculation in the Early 20th Century: Punched-Card Machinery in Business and Government," *Technology and Culture* 31 (October 1990): 753–79; Martin Campbell-Kelly, "Large-Scale Data Processing in the Prudential, 1850–1930," *Accounting, Business and Financial History* 1 (1992): 117–39; JoAnne Yates, "Co-evolution of Information-Processing Technology and Use: Interaction between the Life Insurance and Tabulating Industries," *Business History Review* 67 (spring 1993): 1–51; Martin Campbell-Kelly, "The Railway Clearing House and Victorian Data Processing," in *Information Acumen: The Understanding and Use of Knowledge in Modern Business*, ed. Lisa Bud-Frierman (London, 1994), 51–74; Martin Campbell-Kelly, "Data Processing and Technological Change: The Post Office Savings Bank, 1861–1930," *Technology and Culture* 39 (1998): 1–32.

16. Stan Augarten, *Bit by Bit: An Illustrated History of Computers* (New York, 1984); Adrian Forty, *Objects of Desire: Design and Society, 1750–1980* (London, 1986); Andreas Knie, 'Generierung und Härtung' technischen Wissens: Die Entstehung der mechanischen Schreibmaschine," *Technikgeschichte* 58 (1991): 101–26; Rodney Dale and Rebecca Weaver, *Machines in the Office* (London, 1993); Ellen Lupton, *Mechanical Brides: Women and Machines from Home to Office* (New York, 1993); Charles W. Wootton and Carel M. Wolk, "The Evolution and Acceptance of the Loose-Leaf Accounting System, 1885– 1935," *Technology and Culture* 41 (2000): 80–98.

17. JoAnne Yates, *Control through Communication: The Rise of System in American Management* (Baltimore, 1989).

18. Jan A. de Jonge, *De industrialisatie in Nederland tussen 1850 en 1914* (Nijmegen, 1976), 234–36; Jan Luiten van Zanden, ed., *The Economic Development of the Netherlands since 1870* (Cheltenham, 1996).

19. Circular letter of Fred. Stieltjes and Company and accompanying testimonials of Remington typewriter users, 18 April 1890, Gemeentearchief Amsterdam, bibliotheek, N.42.175.18, nos. 2 and 3.

20. Cornelia C. van de Graft, *Lotgevallen van een Amsterdams koopmanshuis 1749–1949* (Amsterdam, 1949), 208.

21. Jensen (n. 14 above); Strom (n. 14 above), 48–49, 62, 177; Francisca de Haan, *Sekse op kantoor: Over vrouwelijkheid, mannelijkheid en macht,* Nederland 1860–1940 (Hilversum, 1992), 141–50.

22. Although the decision concerning the purchase of a typewriter undoubtedly was the office manager's, at some companies that already used one or more typewriters, employees insisted upon the purchase of additional machines; B. P. A. Gales, *Werken aan zekerheid: Een terugblik over de schouder van AEGON op twee eeuwen verzkeringsgeschiedenis* (The Hague, 1986), 178.

23. Yates, *Control through Communication* (n. 17 above), 39–45.

24. On the invention of the Dictaphone, see Edward J. Pershey, "Drawing as a Means to Inventing: Edison and the Invention of the Phonograph," in *Working at Invention: Thomas A. Edison and the Menlo Park Experience,* ed. William S. Pretzer (Dearborn, Mich., 1989), 100–15; Mark Clark, "Suppressing Innovation: Bell Telephone and Magnetic Recording," *Technology and Culture* 34 (1993): 521–37; David Morton, *Off the Record: The Technology and Culture of Sound Recording in America* (New Brunswick, N.J., 2000).

25. De Haan, *Sekse op kantoor,* 45, 139.

26. Yates, *Control through Communication,* 46–50.

27. Various advertisements, catalogues, and price-lists of typewriter companies, Gemeentearchief Amsterdam, bibliotheek, N.41.151.28-36 and N.42.175.10–18.

28. Circular letter, Amsterdam Bar-Lock Company, circa 1905, Gemeentearchief Amsterdam, bibliotheek, N.41.151.29, no. 1. In practice, the number of readable copies that could be made was probably less than ten.

29. Circular letter, Lutkie & Smit, Amsterdam, August 1904, Gemeentearchief Amsterdam, bibliotheek, N.42.175.11, no. 5.

30. Gemeentearchief Utrecht, Secretariearchief VI 11318, overzicht jan. 1929 van bij gemeentelijke afdelingen en diensten in gebruik zijnde type- en kantoormachines; Gales (n. 22 above), 178–79.

31. De Wit and Van den Ende, "The Emergence of a New Regime" (n. 13 above).

32. For the history of the MABO, see Commissie Tak, *Rapport over de oorzaken van de ontwrichting en van de verergerde ontwrichting van den Postcheque- en Girodienst en de schuldigen daaraan* (The Hague, 1924).

33. Peter Hellema and Joop Marsman, *De organisatie-adviseur: Opkomst en groei van een nieuw vak in Nederland 1920–1960* (Amsterdam, 1997).

34. Jan van Oorschot, "De ontwikkeling van de kantoormachines en de taak van de Rijkskantoormachinecentrale," *Informatie* 20 (1978): 718–24.

35. Peter Zwaal, *Een gemene zaak: Efficiency-ingenieurs, accountants en psychotechnici & en het Nederlandsch Instituut voor Efficiency als presentatieplatform van scientific management 1925–1940* (Rotterdam, 1991); Erik S. A. Bloemen, *Scientific Management in Nederland 1900–1930* (Amsterdam, 1988), 165–81; Cita Hartveld, *Moderne zakelijkheid: Efficiency in wonen en werken in Nederland 1918–1940* (Amsterdam, 1994), 216–34.

36. A. M. J. Kruissink, *Een en ander over de "STUMOKA"* (Purmerend, 1931).

37. R. G. ter Haak, *Kantoormachines en administratiesystemen: Een beschrijving van moderne hulpmiddelen ten dienste van de administratie* (Amsterdam, 1927), 278–304.

38. A. Meeuwis, *Hulpmiddelen der administratieve techniek* (The Hague, 1959), 3–4.

39. In 1919, the Dutch Association of Stationers (Nederlandsche Vereeniging van Kantoorboekhandelaren, NEVEKA) founded its own monthly magazine, and the Association of Importers and Producers of Office Machines (Vereeniging van Importeurs en Fabrikanten van Kantoormachines, VIFKA) regularly organized exhibitions on office efficiency from 1924 onward.

40. Jozef L. de Jager, *De bank van de gulden: Organisatie en personeel van De Nederlandsche Bank 1814–1989* (Amsterdam, 1989), 95, 121.

41. Joan I. W. Klaassen, "Adresseermachines: Ontwikkeling en gebruik van adresseermachines in de administratie" (graduation thesis, Technical University Eindhoven, 1994), 56–57.

42. Klaassen, 70–75.

43. Klaassen, 41.

44. On punched-card technology, see Martin Campbell-Kelly, "Punched-Card Machinery," in *Computing before Computers*, ed. William Aspray (Ames, Iowa, 1990), 122–55; Norberg (n. 15 above).

45. James Connolly, "History of Computing in Europe," unpublished report, IBM World Trade Corp., 1968, 19–21.

46. Jan van den Ende, *Knopen, kaarten en chips: De geschiedenis van de automatisering bij het Centraal Bureau voor de Statistiek* (Voorburg and Heerlen, 1991), 2–28.

47. Connolly, 20, E-3/7.

48. The transfer of ideas and punched-card technology from the Rotterdamsche Bank to other companies is dealt with in De Wit and Van den Ende, "The Emergence of a New Regime" (n. 13 above).

49. ABN AMRO, Historical Archive, Amsterdam, Archive Rotterdamsche Bank, no. 109, verslag vergadering Raad van Commissarissen, 16 November 1921.

50. From about 1920 onward, the bank used both Hollerith and Powers machinery. In 1925 and 1930 more Hollerith punched-card machinery was ordered; Archive Rotterdamsche Bank, no. 94, stukken betreffende The Tabulating Machine Company.

51. De tentoonstelling op het gebied van de openbare en particuliere bedrijfsadministratie T.O.P.A. 1926, 3 vols. (Purmerend, 1926–27).

52. Sam Roet, "Moderne expeditie- en portiadministratie," Administratieve Arbeid (1924), 84–92; H. Fedder, "Personeels-Administratie," Administratieve Arbeid (1927), 186–90.

53. At the Dutch Central Statistical Bureau in 1937, for example, outside consultants analyzed the flow of forms and punched cards in the Trade Statistics Department, through different units using machinery that included sorting devices, punchers, punched-card sorters, tabulators, calculating machines, typewriters, archival devices. Van den Ende, *Knopen, kaarten en chips* (n. 46 above), 49–52.

54. Yates, "Co-evolution of Information-Processing Technology and Use" (n. 15 above); Campbell-Kelly, "Punched-Card Machinery" (n. 44 above), 142; *De tentoonstelling op het gebied van de openbare en particulierebedrijfsadministratie* , 2:310–32; R. G. ter Haak, "Powers en haar nieuwe vindingen," *Administratieve Arbeid* (1931), 287–320, 345–66.

55. Ter Haak, "Powers en haar nieuwe vindingen"; M. C. Winterstein, "De alfabetische Hollerith-Tabelleermachine," *Administratieve Arbeid* (1932), 32–34. See also James W. Cortada, *Before the Computer: IBM, NCR, Burroughs, and Remington Rand and the Industry They Created, 1865–1956* (Princeton, N.J., 1993), 106–9.

56. Catalogus "Efficiency" *Tentoonstelling van moderne hulpmiddelen voor fabrieks- en kantooradministratie van 18 tot en met 27 maart 1926* (Amsterdam, 1926); *Administratieve Arbeid* (1926), 106–7.

57. C. W. Wintersteijn, *Banken, hare boekhouding en organisatie*, vol. 1 (Leiden, 1932), 50–53.

58. Connolly (n. 45 above), 20.

59. Meeuwis (n. 38 above), 198–218; Studiegroep Industrie Den Haag, *Administratieve techniek* (The Hague, 1954).

60. *Modern Kantoor*, July–August 1958, 121; VIFKA Archive, Woerden, files Efficiency Exhibitions 1959–1970, Efficiency Beurs 1960.

61. D. Pietersma, "Enkele kleine integratiehulpmiddelen in de administratie," in *Omschakeling naar automatie: Enkele bijdragen over de ontwikkeling naar automatische informatieverwerking*, ed. N. V. Philips (Eindhoven, 1960), 10.0–6.

62. A. Meeuwis, "Aanpak van automatie in de administratie," in Philips, 2.0–11. The telex service of the Dutch Post, Telegraph and Telephone Administration was using punched paper tape as a transmission medium by the early 1950s.

63. By 1968 almost a thousand computers had been installed in the Netherlands. The Association of Importers and Producers of Office Machines estimated that eighty percent of these were used for administrative purposes: R. J. Romein, "Computers in Nederland," *Informatie* 11 (1969): 316; *Kantoor en Efficiency*, September 1968, 4011. The IBM 650 was called the Model T of computers; Rick Szostak, *Technological Innovation and the Great Depression* (Boulder, Colo., 1995), 198.

64. Speech, H. J. A. Hendrikx, 28 March 1957, and annual reports, 1947–57, Rijksarchief Gelderland, Arnhem, archief Heidemaatschappij, no. 538.

65. Paul Ceruzzi, "Electronics Technology and Computer Science, 1940–1975: A Coevolution," *Annals of the History of Computing* 10 (1989): 261.

66. Paul Ceruzzi, "Crossing the Divide: Architectural Issues and the Emergence of the Stored Program Computer, 1935–1955," *IEEE Annals of the History of Computing* 19 (1997): 5–12.

67. Punched-card machines were programmed by wired switchboards that connected the reading devices of the machines with tabulating and print mechanisms, as opposed to putting abstract code into computer memory.

68. Paul N. Edwards, "Y2K: Millennial Reflections on Computers as Infrastructure," *History and Technology* 15 (1998): 7–29, at 13; Martin Campbell-Kelly and William Aspray, *Computer: A History of the Information Machine* (New York, 1996), 105–30.

69. Ellen van Oost, "Administratieve automatisering," in *Techniek in Nederland in de twintigste eeuw*, ed. Johan Schot et al. vol. 1 (Zutphen, 1998), 326–27.

70. The SSAA was not a commercial consulting firm or a publicly financed nonprofit organization. During the first ten years its income came mainly from the contributions of associated companies that pioneered automation. From the end of the 1960s onward, it was gradually transformed into a training and examination institute that regulated and controlled the education of the new computer professionals. For the history of the SSAA, see Ellen van Oost, *Nieuwe functies, nieuwe verschillen: Genderprocessen in de constructie van de nieuwe automatiseringsfuncties 1955–1970* (Delft, 1994), 209–49.

71. Algemeen Nederlands Persbureau (General Dutch Press Association) telex news message, archive of the Rijkscentrale voor Mechnische Administratie, Apeldoorn, 21 June 1963.

72. The technical development of computers usually is characterized in generations. The first generation (1953–58) used vacuum tubes as central elements in the processing unit. The second generation (1958–64) used transistors. Transistorized computers were smaller and faster than those with vacuum tubes and did not need to be cooled. The third generation (1964–1972) incorporated the new integrated circuit (IC) technology. Not only were these machines (again) faster and smaller, but their internal programming was organized differently. They had a separate operating system, making such new functionalities as time-sharing and multitasking possible.

73. *Kantoor en Efficiency*, September 1968, 4011.

74. Meeuwis, *Hulpmiddelen der administratieve techniek* (n. 38 above), 6.

75. S. Swaab, *Methodiek van de automatische administratie* (Alphen aan den Rijn, 1962), 174.

76. Van Oost, *Nieuwe functies, nieuwe verschillen* (n. 70 above), 175–97. In general, the 1950s and 1960s were a period of persistent labor shortage in the Netherlands; Van Zanden (n. 18 above).

77. In the mid-1960s the R&D department of the governmental Postal, Telegraph and Telephone Services developed the ALPINO system, a machine that could automatically read and punched numerical information, in an attempt to solve the data-entry problem at the Dutch Giro Service; Van Oost, *Nieuwe functies, nieuwe verschillen*, 195. Mark sensing was successfully used by the Dutch Central Statistical Office for the 1971 census: Van den Ende, *Knopen, kaarten en chips* (n. 46 above), 96.

78. Because of labor shortages, the Giro Service, where data processing was still predominantly done by hand until the end of the 1950s, twice had to stop accepting new accounts. After the second time, in 1961, management decided to initiate an automation project. Interestingly, the Giro Service then became one of the leaders in the field of office automation: Van Oost, *Nieuwe functies, nieuwe verschillen*, 95–133.

79. Meeuwis, Hulpmiddelen der administratieve techniek, 284.

80. The Giro Service called this external integration: G. F. J. A. Groen, "De taaie ponskaart," in Een halve eeuw Postcheque- en Girodienst, ed. Hendrik Reinoud et al. (Utrecht, 1968), 236–42, at 239.

81. Meeuwis, Hulpmiddelen der administratieve techniek (n. 38 above).

82. Paul Atkinson, "Computer Memories: The History of Computer Form," History and Technology 15 (1998): 89–120.

83. Atkinson, 107.

84. The miniaturization that eventually led to minicomputers and personal computers stimulated a break in the centralized organization of data processing. Their rapid adoption in the 1980s resulted in a wide array of "incompatible, stand-alone machines and software." Managers attempting to regain (central) control were partly responsible for the rise of networked systems; Edwards (n. 68 above), 22.

Afterword

THOMAS P. HUGHES

Fix can mean "fix up" as in "fix up a car," but it can also mean fix as "in a fix," in the sense of difficulty. Most of the papers at the Hagley Conference on technological fixes suggested that technological fixes leave us in a fix. For example, Shelley McKellar found early artificial hearts left patients with unintended negative consequences that can be subsumed under the rubric *quality of life*. Timothy J. LeCain found that a technological fix that seemed to solve a smelter's arsenic and sulfur release problem left unintended consequences that he labels transformational, relocational, and delayed, all of which, to use Edward Tenner's words, "bite back."

Technological fixes are partial, reductionist responses to complex problems. They are not solutions. Yet Americans have avidly embraced technological fixes. The reason, in part, is their long-standing technological enthusiasm. Enthusiasm can be defined as fanatic religious ardor or supposed possession by a god. To Perry Miller, a Harvard historian and professor of literature, nineteenth-century Americans seemed possessed by a fanatic religious ardor or by a god as they flung "themselves into the technological torrent," shouting "with glee in the midst of the cataract," and crying "to each other as they went headlong down the chute that here was their destiny."[1]

In short, a god named technology has possessed Americans. Traditionally a god controls some important aspect of human life. Greek and Roman gods divided the responsibility, but Christians and Jews have preferred one god in control, or at least designing and putting the softly deterministic clockwork into motion. Even before Americans found technology a god, Oswald Spengler in *Decline of the West* (1918–22) spotted an early turn to technology as god when he found medieval monks searching out the secrets of the universe in order to control its workings. Later, this search was called *scientific* and its application, *technology*. Spengler did not let the monks off easily. He accused them of the fatal sin of pride for challenging God's authority. Today Americans are guilty of scientific and technological pride; foreign critics prefer to call them arrogant.

As the twentieth century opened, Americans believed that science-based technology empowered them to control and to alter the physical world according to their own blueprint, not God's design. Thomas Edison and the independent inventors of intuitive genius had opened a cornucopia of material goodies, and scientists in industrial research laboratories promised that they did not rely on slippery intuition, but upon a scientific method. This method not only sustained the continued mass production of consumer goods, but also allowed humans to manipulate the natural environment.

Americans' faith in technology in the twentieth and twenty-first centuries is analogous to the faith that medieval people had in God. God, they believed, not only controlled them and nature, but also occasionally performed miracles. Their religious ardor sustained their faith in God and miracles. Similarly, Americans' faith in technology supports their passion for the miracle of technological fixes. Miraculous technological fixes today occur mainly in the military realm rather than the civil. During the interval between the two world wars, large manufacturing firms gave American technology a civil image. Foreigners associated America with the democratization of consumer goods, with economic democracy. Today, however, the image of American technology is a formidable military one.

Western medieval crusaders invaded the Middle East with the cross held high before them. They had confidence in a religious fix. Recently American forces invaded Iraq confident that their technology would prevail. Defense Secretary Donald Rumsfeld had faith that a relatively small force of soldiers, as compared to the Gulf War of 1991, supported by lethal technology, would overwhelm the technologically backward Iraqi army in short order, bring a regime change, and democracy.

The military success in Iraq is clearly a partial, reductionist technological fix. Although it brought a military victory, it will not solve the political, social, and economic problems of a devastated nation. There may be a technological fix for the Iraqi petroleum industry. Conceivably media technology, including an Iraqi Internet, could help foster a democratic spirit. Technology alone will not bring democracy.

Democracy is culture based and culture embraces many more components than technology. Culture at its deepest level includes fundamental ideas about human relations, the use of force, the ways in which differences are resolved, responsibilities of the haves for the have-nots, and so on.

In general, technological fixes fail to take a systems approach to the complex cultures or systems in which problems are embedded. Two of the papers in this volume are not about technological fixes because the problem solvers took a systems approach. Shane Hamilton in "Long-Haul Trucking and the Technopolitics of Industrial Agriculture, 1945–1975" argues persuasively that the U.S. Department of Agriculture (USDA) helped counter the rising cost of food for consumers and at the same time upheld the income of farmers for their products. The USDA did this by introducing a system to reduce the distribution cost of food, not simply a technological fix. It stimulated and presided over the introduction of a system involving highly mechanized farming, trucks with freezer facilities, frozen food processing plants, and suburban supermarkets.

Similarly, Onno de Wit, Jan Van den Ende, Johan Schot, and Ellen van Oost in "Innovations Junctions: Office Technologies in the Netherlands, 1880–1980" write the history of an office *innovation junction*, which they define as the systematic integration of heterogeneous technologies, such as duplicating, accounting, bookkeeping, tabulating, and addressing, to solve office problems and optimize administrative procedures,

Finally, Paul Ceruzzi in "The 'Problem' of Computer-Computer Communication, 1995–2000: A Technological Fix?" brings to the fore another aspect of technological fixes — unintended consequences. While technological fixes ignore the systematic interactions required to solve a complex problem, they often initiate unintended consequences, as well. Ceruzzi tells how the effort to establish computer-computer communication led to a set of unintended sequences including the ARPAnet, World Wide Web, and browsers. The commercialization of the Internet became the greatest unintended consequence for the inventors and developers of the computer-computer communication. They intended the Internet to be used for academic and government research and to lower costs.

From the Hagley conference on technological fixes, we should conclude that technology is deeply bound up in a cultural matrix. To deploy technology to solve problems, therefore, problem solvers should take a systems approach that deploys technology along with other components that respond to the complexity of the problem to be solved.

Note

1. Perry Miller, "The Responsibility of Mind in a Civilization of Machines," *The American Scholar*, XXXI (Winter 1961–1962) pp. 54–55.

Contributors

Michael Ackerman is a Ph.D. candidate in American history at the University of Virginia, Charlottesville. His forthcoming dissertation will explore the contested interpretations of the newer knowledge of nutrition, and the influence of nutrition science on the rise of the modern health foods movement in the middle third of the twentieth century.

Warren Belasco teaches American studies at the University of Maryland, Baltimore County. He is the author of *Appetite for Change: How the Counterculture Took on the Food Industry* (Cornell University Press, 1993) and *Meals to Come: A History of the Future of Food* (University of California Press, forthcoming). He is on the editorial board of *Gastronomica, Food and Foodways* and *The Oxford Encyclopedia of American Food*.

Paul E. Ceruzzi is curator of Aerospace Electronics and Computing at the Smithsonian Institution's National Air and Space Museum (NASM) in Washington, DC. His work there includes research, writing, planning exhibits, collecting artifacts, and lecturing on the subjects of microelectronics, computing, and control as they apply to the practice of air and space flight. He is the author or coauthor of several books on the history of computing and related topics: *Reckoners: The Prehistory of The Digital Computer (1983); Smithsonian Landmarks in the History of Digital Computing* (1994, with Peggy Kidwell); *A History of Modern Computing* (1998, second edition 2003); and *Beyond the Limits: Flight Enters the Computer Age* (1989). He is currently working on a history of systems integration firms located in the Washington, DC region. Dr. Ceruzzi has curated or assisted in the mounting of several exhibitions at NASM, including "Beyond the Limits: Flight Enters the Computer Age," "The Global Positioning System: A New Constellation," "Space Race," and "How Things Fly."

James R. Fleming is a historian of science and technology and professor of Science, Technology and Society (STS) at Colby College, Waterville, Maine. His teaching bridges the sciences and the humanities, and his research interests involve the history of the geophysical sciences, especially meteorology, climatology, and oceanography. His books include *Meteorology in America, 1800-1870* (Johns Hopkins University Press, 1990, 2000); *Historical Perspectives on Climate Changes* (Oxford University Press, 1998); *Science, Technology, and the Environment: Multidisciplinary Perspectives* (Akron University Press, 1994); *Weathering the Storm: Sverre Petterssen, the D-Day Forecast and the Rise of Modern Meteorology* (American Meteorological Society, 2001); *International Perspectives on the History of Meteorology: Science and Cultural Diversity* (Mexico City: UNAM, in press). Professor Fleming is also president of the International Commission on History of Meteorology, program officer of the History of Earth Sciences Society, history editor of the Bulletin of the American Meteorological Society, advisory editor of the Papers of Joseph Henry, and member of the Electorate Nominating Committee of the American Association for the Advancement of Science (AAAS).

Shane Hamilton is a doctoral candidate in the history and social studies of science and technology at Massachusetts Institute of Technology. His doctoral dissertation, "Trucking Country: Food, Farms, and Freight in America's Rural Industrial Landscape, 1945-1975," explores the political economy and culture of long-haul trucking in modern industrial agriculture. The Agricultural History Society recently awarded him the Edward E. Everetts prize for "Cold Capitalism: The Political Ecology of Frozen Concentrated Orange Juice," forthcoming in *Agricultural History*.

Thomas P. Hughes is Mellon Professor Emeritus at the University of Pennsylvania and Distinguished Visiting Professor at the Massachusetts Institute of Technology.

Timothy LeCain is an assistant professor of history at Montana State University in Bozeman. He has published several recent articles on modern mining technology and the environment and is currently revising his book manuscript on these topics, "Moving Mountains: Technology and the Environment in Western Copper Mining."

Shelley McKellar is assistant professor in the Department of History and History of Medicine Program at the University of Western Ontario, where she teaches courses in the history of health and medicine, medical technology, women's health, and conceptualization of the body. She is author of *Surgical Limits: The Life of Gordon Murray* (University of Toronto Press, 2003).

Ellen van Oost is associate professor in the faculty of Philosophy and Social Sciences of the University of Twente, the Netherlands.

Carolyn Thomas de la Peña is assistant professor of American Studies at the University of California, Davis, where she also teaches in the program for Technocultural Studies. She is the author of *The Body Electric: How Strange Machines Built the Modern American* (New York University Press, 2003) and is currently working on projects exploring technologies in consumer spaces and the American fascination with techno-foods.

Lisa Rosner is professor of history at the Richard Stockton College of New Jersey. Her publications include *Medical Education in the Age of Improvement: Edinburgh Students and Apprentices 1760-1826* (Edinburgh University Press, 1991), *The Scandalous Life of Alexander Lesassier* (University of Pennsylvania Press, 1999), and (as consulting editor) *Chronologies of Science: From Stonehenge to the Human Genome Project* (ABC-CLIO, 2002).

Johan Schot is professor of the history of technology at the Eindhoven University of Technology and the University of Twente, both in the Netherlands.

Jim Tobias, president of Inclusive Technologies, has worked in the field of technology and disability for twenty-five years. Beginning at Berkeley's Center for Independent Living, he has worked as a rehabilitation engineer with schools, hospitals, private organizations, companies, and state and federal agencies. He worked for ten years at Bell Labs and Bellcore, providing telecommunications and disability consulting for Bell companies and other telecommunications and disability information industry clients, before leaving to found Inclusive Technologies. His technical background supports Inclusive Technologies' hardware and software services. In addition, he specializes in accessible business practices: primary and secondary market research and analysis, customer surveys, focus groups, product trials, product management, strategic partnership development, staff training, internal team-building, and consumer and other stakeholder liaison.

Frank Uekoetter, born in 1970, studied history, political science, and the social sciences at the universities of Freiburg and Bielefeld in Germany and the Johns Hopkins University in Baltimore, Maryland. In 2001 he completed a Ph.D. on the history of air pollution control in Germany and the United States. He organized a conference on "nature protection in Nazi Germany" in 2002 under the auspices of the German minister for the environment, Juergen Trittin. He is currently starting a new project on the history of agricultural expertise in the twentieth century.

Jan Van den Ende is associate professor, Management of Innovation Unit of the Rotterdam School of Management.

Onno de Wit is senior researcher, Management of Innovation Unit of the Rotterdam School of Management.

Index